AF811624

Mathematical Models for Dental Materials Research

Alex Fok • Hooi Pin Chew

Mathematical Models for Dental Materials Research

Springer

Alex Fok
Department of Restorative Sciences
University of Minnesota
Minneapolis, MN, USA

Hooi Pin Chew
Department of Restorative Sciences
University of Minnesota
Minneapolis, MN, USA

ISBN 978-3-030-37848-6 ISBN 978-3-030-37849-3 (eBook)
https://doi.org/10.1007/978-3-030-37849-3

© Springer Nature Switzerland AG 2020
This work is subject to copyright. All rights are reserved by the Publisher, whether the whole or part of the material is concerned, specifically the rights of translation, reprinting, reuse of illustrations, recitation, broadcasting, reproduction on microfilms or in any other physical way, and transmission or information storage and retrieval, electronic adaptation, computer software, or by similar or dissimilar methodology now known or hereafter developed.
The use of general descriptive names, registered names, trademarks, service marks, etc. in this publication does not imply, even in the absence of a specific statement, that such names are exempt from the relevant protective laws and regulations and therefore free for general use.
The publisher, the authors, and the editors are safe to assume that the advice and information in this book are believed to be true and accurate at the date of publication. Neither the publisher nor the authors or the editors give a warranty, expressed or implied, with respect to the material contained herein or for any errors or omissions that may have been made. The publisher remains neutral with regard to jurisdictional claims in published maps and institutional affiliations.

This Springer imprint is published by the registered company Springer Nature Switzerland AG
The registered company address is: Gewerbestrasse 11, 6330 Cham, Switzerland

Foreword: Enjoying and Applying Mathematics in Dental Materials Research

> I had the feeling that, through the surface of atomic phenomena, I was looking at a strangely beautiful interior, and felt almost giddy at the thought that I now had to probe this wealth of mathematical structures nature had so generously spread out before me. . . .
>
> Werner Heisenberg, quantum physicist

I am truly glad to write this Foreword to *Mathematical Models for Dental Materials Research.* Writing gives me the opportunity to say something more general about mathematics as well as to endorse the content produced by two highly competent colleagues: Dr. Alex Fok and Dr. Hooi Pin Chew.

Mathematics has played a significant role throughout human history being closely intertwined with science, engineering and technology, plus its intellectual ties with philosophy and theology. Mathematics is for all. The popularity of Sudoku is evidence that a wide range of people can enjoy mathematics. Unfortunately, many intelligent people have had bad experiences with their mathematical education. This was my own experience during much of my high school education. What passed for teaching largely reduced to repetitive exercises and attempting to solve problems with insufficient explanation of the underlying theory or rationale. It was only when I went to university, studying mathematics alongside major courses in physics and chemistry, that light dawned! Professional mathematicians at university taught the mathematics courses. They developed the underlying theories systematically and logically. I flourished under that regime and continued to seek out mathematical treatments of phenomena in physics, chemistry and materials science. My most successful research paper ever, from my Ph.D., would have been impossible without good mathematical understanding.

I began to see that mathematics is not only understandable and useful but also intellectually rewarding and beautiful. Especially when freed from the artificial constraints of examinations and competition, mathematics can be an engaging and intellectually satisfying pastime, with unexpected relevance to normal life [1].

My philosophy—or theology—of science is based on the concept of *theistic creation.* This embodies the view that Mind is antecedent to Matter and that created reality has a real existence susceptible to exploration and discovery by human beings

[2]. A popular alternative view (*reductive materialism*) envisages mind as being merely a highly organized state of matter that has arisen by purposeless processes or chance. It is questionable whether this view is internally coherent, as has been recognized recently by the leading philosopher, Thomas Nagel [3]. A corollary of that view is that mathematical "structures" are *merely* inventions of the human mind imposing order on a fundamentally chaotic universe. The theistic view, akin to philosophical *idealism*, holds that mathematical structures exist "out there" and await *both* human discovery *and* important elements of human development/invention [4]. This would explain why appropriate mathematical models are capable of forming such an excellent "fit" to physical reality. The Nobel Prize winner, Eugene Wigner, expressed his astonishment at the "Unreasonable effectiveness" of mathematics in the natural sciences [5].

Another key feature of mathematics is its history—at least over the past 3000 years. There are numerous books on the history of mathematics, ranging from the boring to the brilliant. Supplementing these is a superb website maintained by the University of St Andrews: the *MacTutor History of Mathematics archive*: (http://www-history.mcs.st-and.ac.uk/). This brings out the essential human dimension and the astonishing achievements of mathematicians such as Leonhard Euler (1707–1783), the Swiss mathematician, physicist, astronomer, geographer, logician and engineer who is now largely unknown to the general public. Euler discovered what is widely regarded as the most beautiful equation in mathematics:

$$e^{i\pi} + 1 = 0$$

This connects the five *fundamental constants* of mathematics (e, π, i, 0, 1), where the imaginary number i is the square root of -1. It includes the three most important *mathematical operations* (addition, multiplication and exponentiation) and includes the constants symbolizing *the four major branches of mathematics*: arithmetic (0,1), algebra (i), geometry (π) and analysis (e). If you have an enquiring mind—the greatest prerequisite for success as a researcher—I hope you will think that this is intriguing and worthy of further investigation.

There are several accessible books that can stimulate and motivate an interest in mathematics. David Acheson—a mathematician at Oxford University—wrote *1089 and All That* [6]—said by a reviewer to be "possibly the nicest maths book ever written"! A colleague at Manchester University, Paul Glendinning, wrote *Maths in Minutes: 200 key concepts explained in an instant* [7]. Each section includes an informative picture and a maximum 200-word explanation. Tony Crilly has written a similar sort of book on *50 Mathematical Ideas* incorporating more extended discussions, without it reading like a formal textbook [8]. Several other books focus upon famous equations in science, explaining their background and significance [9–11].

Researchers in Dental Materials fall into two main classes. Firstly, there are qualified clinicians whose mathematical studies may be a distant memory. Secondly there are scientists and engineers who have had a good mathematical education and now face the challenge of understanding the biological and clinical context of their work. The latter group should already have sufficient background to understand and profit from the present volume. Nevertheless a compact volume of Mathematical Terms and equations can be a convenient desk reference [12].

Members of the first group—the clinicians—need encouragement and guidance to re-connect with mathematics, so that they also can appreciate and enjoy mathematical modelling in dental research. Their situation may parallel the needs of, for example, beginning university students of chemistry who may lack sufficient mathematical preparation. Several books have been written specifically for such persons and they are highly suitable for researchers in dental materials [13–15].

Some of the basic ideas that underlie the use of mathematics by scientists concern equations: especially linear equations. Chapter 1 of the book by Tebbutt [14] is clear, concise and helpful. Monk and Munro [15] cover the same ground in their first 13 chapters with more detail and examples of chemical calculations. Scientists must be able to use, rearrange and manipulate equations, turning them round to get the information required. It is necessary to plot data in the manner that an equation indicates is correct or if necessary change an equation so that it is in an appropriate form. Experimentalists must be able to use data, plot and correctly label graphs, derive parameters from the slope or the intercept and present the result with the correct units. It should be checked that not only is an equation balanced properly but that the units also balance by being dimensionally correct. Moreover one should know how to convert units from one system to another.

Among the most useful mathematical functions are the inverse functions: Logarithms and Exponentials, including "stretched" or fractional power exponentials. They are useful because they so closely mirror the way that the physical, material and biological worlds change over distance and/or time, in many respects. Examples include the increasing attenuation (by absorption and/or scattering) of light as it passes through solutions or solids (the Beer–Lambert law), the kinetics of many chemical reactions, including setting processes in dental materials, diffusion processes, the rates of molecular relaxation processes—including mechanical stress relaxation—within condensed matter, the rates of radioactive decay and the growth of colonies of bacteria. Another instance occurs with the Boltzmann distribution of molecular energies, which underlies the Arrhenius equation describing the effect of temperature on the rates of chemical reactions.

Gottfried Leibnitz (1646–1716) and Isaac Newton (1642–1727) were contemporary *natural philosophers* that independently developed the differential and integral calculus. These mathematical disciplines continue to have an enormous impact upon science and engineering. Related to both is the subject of *differential equations* that feature in Chaps. 3–6, 8, and 9 of the present volume. It is often possible to "write down" a differential equation to describe a given situation. For example, consider filling a bathtub with water from a tap while the plughole is open. Assuming that water is neither created nor destroyed in the process (i.e. conservation), then the volume [$V(t)$] of water in the tank at time t will depend upon the *rates of inflow* and *outflow* plus some initial (or *boundary*) conditions. "Solving" the resultant differential equation means finding an expression for one or more quantities of interest such as the height of water in the bath.

From what we have said so far, one of the main practical benefits of applying mathematical tools in dental materials research is to *compactly quantify* behaviour of materials in their host environments. This can facilitate prediction of behaviour over—for example—longer time scales. Sometimes—or even often—such predictions fail. But the fact that we had a prediction means that we had a rational theory and that such a theory can be modified for future application if we can analyse discrepancies between predictions and actual outcomes.

Most of science, and especially materials science, involves model building. Mathematical models are key components of such an approach. Today they merge with computer modelling in its various forms, including *finite element* modelling and *numerical solutions* of complex differential equations. The great strengths of this volume are the real world examples of dental materials research in the successive chapters. In turn, this is an outcome of the outstanding expertise of both authors.

I have known Professor Alex Fok for many years since his time as a senior academic in the Engineering School at the University of Manchester, UK. He moved gradually into dental materials research bringing his engineering and mathematical acumen plus knowledge of areas such as nuclear graphite materials that have significant parallels to dental composites. As Director of the *Minnesota Dental Research Center for Biomaterials and Biomechanics* (MDRCBB) since 2007, Dr. Fok is renowned internationally for his ability to instil more rigorous engineering principles into the design and assessment of dental restorations and treatments so as to improve their reliability. Dr. Hooi Pin Chew is a distinguished clinician, with a Ph. D. from the University of Manchester. She has made major contributions to research on dental erosion and dental caries using advanced techniques such as light-induced fluorescence and optical coherence tomography. This unique volume is a testament to their joint vision and accomplishments.

I warmly commend this book to a worldwide audience among the dental biomaterials community.

School of Medical Sciences and Photon Science Institute, University of Manchester, Manchester, UK

David C. Watts

References

1. Adam, J.A.: *A Mathematical Nature Walk*. Princeton University Press, Princeton (2009)
2. Watkin, C.: *Thinking Through Creation*. P&R, Phillipsburg (2017)
3. Nagel, T.: *Mind and Cosmos: Why the Materialist Neo-Darwinian Conception of Nature Is Almost Certainly False*. Oxford University Press, Oxford (2012)
4. Polkinghorne, J. (ed.): *Meaning in Mathematics*. Oxford University Press, Oxford (2011)

5. Wigner, E.: *The unreasonable effectiveness of mathematics in the natural sciences*. Commun. Pure Appl. Math. **13**, 1–14 (1960)
6. Acheson, D.: *1089 and All That: A Journey into Mathematics*. Oxford University Press, Oxford (2002)
7. Glendinning, P.: *Maths in Minutes*. Quercus Books, London (2011)
8. Crilly, T.: *50 Mathematical Ideas*. Quercus Books, London (2007)
9. Farmelo, G.: *It Must Be Beautiful: Great Equations of Modern Science*. Granta, London (2002)
10. Crease, R.P.: *The Great Equations*. WW Norton, New York (2008)
11. Stewart, I.: *17 Equations that Changed the World*. Profile Books, London (2012)
12. Downing, D.: *Dictionary of Mathematical Terms*. Barron's Educational Series, New York (2009)
13. Tebbutt, P.: *Basic Mathematics for Chemists*. Wiley, Chichester (2001)
14. Cockett, M.C.R., Doggett, G.: *Maths for Chemists: Volumes 1 & 2*. Royal Society of Chemistry, London (2003)
15. Monk, P., Munro, L.J.: *Maths for Chemistry: A Chemist's Toolkit of Calculations*. Oxford University Press, Oxford (2010)

Preface

This book was borne out of a desire to address an imbalance in the way dental materials research is being carried out.

A typical paper on dental materials research often goes as follows. A new material or treatment procedure has been invented which has the potential of outperforming existing ones. A series of tests is thus carried out to compare the performance of these materials or procedures. The null hypothesis would be that there are no statistically significant differences in performance among the groups. Statistical analysis such as ANOVA is then carried out to analyse the differences among group means to decide whether the null hypothesis should be rejected. Such statistical analysis allows researchers to arbitrarily set a threshold probability (conventionally 5%) to determine whether the differences among group means occurred due to chance. The utility of such statistical analysis, however, stops here.

Limited by the output of such analysis, the discussion that ensues is often vague and speculative. Authors would propose why or how the results turned out the way they did without being able to provide any quantitative assessment of those factors that could have affected the outcome to support their arguments. Without such an assessment, one would not be able to decide, under the particular conditions, what is important and what is not. For example, some conditions, which are included to make the tests clinically representative, actually have little effect on the outcome. On the other hand, a small change in the protocol, which is thought, wrongly, to be insignificant, could drastically alter the results. Thus, there have been many inconsistent, if not contradictory, results reported by different researchers—sometimes even by the same researchers—on many topics of dental materials research. For example, some papers report that shrinkage stress is proportional to the Young's modulus of a resin composite while no such correlation was found in others. As a result, one is frequently left with the feeling of dissatisfaction, frustration or even confusion after reading such papers.

The aim of this book is to present a more mechanistic approach, i.e. mathematical modelling, for carrying out dental materials research. It introduces to dental materials researchers, especially those without a strong mathematical background, some simple mathematical models and techniques, which would allow them to go beyond the null hypothesis. Through a series of case studies with practical applications drawn from operative dentistry, endodontics, prosthodontics, etc. the reader will be shown how to

1. construct a mathematical model for the behaviour of dental materials by making informed assumptions of the relationship between cause and effect;
2. simplify the model by making appropriate simplifications;
3. calibrate the model by calculating the key parameters using experimental results; and, finally,
4. refine the model when there are discrepancies between predictions and experiments.

Examples include fluorescence measurement of demineralized tooth tissues, fracture of root dentine, shrinkage of resin composites, leaching of fluoride ions, the weakest link theory for brittle materials, polymerization kinetics and natural frequency of vibration of implants. The mathematical models progress from the simple straight-line equation for systematic errors in measurements to nonlinear simultaneous differential equations for the formation of polymeric chains via free radicals. However, only elementary calculus is required to follow the examples, and all the problems can be solved by using MS EXCEL© spreadsheets.

This mechanistic approach for performing dental materials research is no doubt more challenging. Not only does it require more mathematical skills, it also forces the researcher to think deeper and more mechanistically about the problem at hand. But the rewards from solving the problem using this approach are much more satisfying than stopping at showing significant differences in performance among the test materials or procedures. The solution thus obtained is more general, and hence it is predictive for conditions other than those considered in the study; it is more mechanistic, and hence it provides deeper insight into the problem.

For those who already have the mathematical skills, it is hoped that this book would get you interested in the wide-ranging topics of dental materials research. We need your input. It is our belief, and indeed our experience, that the most fruitful results are often produced when clinicians work closely with physical scientists or engineers.

Last but not least, we would like to thank our students and colleagues who have contributed to the case studies used in this book. We also thank Professor Michael Burrow of the University of Hong Kong for his valuable comments on the final draft. All errors, of course, remain ours. Particular thanks go to Professor David Watts of the University of Manchester (UK) and Editor-in-chief of Dental Materials for his insightful foreword. His passion for mathematics and his dedication to dental materials research is an inspiration for us all.

Minneapolis, MN, USA Alex Fok
Hooi Pin Chew

Contents

Chapter 1
Systematic Errors in Longitudinal Measurements

1.1 Introduction

In this opening chapter, we will look at the accuracy of measurement or observation, the most fundamental task in experimental research. In particular, we will consider systematic errors due to temporal changes in the measuring instrument. We will use the all-familiar straight-line equation, $y = mx + c$, to estimate these errors and show how they can be used to correct the measurements.

Very often in dental materials research, certain quantities of the same samples have to be measured at different time points to investigate the temporal changes induced by mechanical, chemical or biological effects. For example, those involving demineralization or remineralization of dental hard tissues require the mineral density to be measured longitudinally. One of the challenges in conducting longitudinal studies is to ensure that the measured changes are from the same region of interest across time points to minimize inter-sample biological variation. The other is to ensure that the changes are caused by the intended effect and not due to confounding conditions such as inconsistency of the measuring instrument.

One of the established methods in tracking mineral density changes of enamel non-destructively is through measuring its surface microhardness [1, 2]. Other non-destructive methods include micro-computed tomography (μCT) [3] or optical methods such as quantitative light-induced fluorescence (QLF) and optical coherence tomography (OCT) [4]. The advantage of using non-destructive methods is that the same region of interest can be measured throughout the test, thus minimizing the sample size and uncertainties due to inherent differences among biological samples. However, for temporal comparison to be accurate, the instrument that we use needs to be able to make consistent measurements over time. For example, the energy of the incident beam of the μCT or OCT machine should stay relatively constant.

© Springer Nature Switzerland AG 2020

A. Fok, H. P. Chew, *Mathematical Models for Dental Materials Research*,
https://doi.org/10.1007/978-3-030-37849-3_1

1.2 Effect of Systematic Errors

Because of biological heterogeneity, there could be large variations in the baseline measurements. In the event that the changes depend on the baseline value, the fractional change of the quantity, *E*, being measured should be considered, i.e.,

$$\frac{\Delta E}{E} = \frac{E(t) - E(0)}{E(0)}, \tag{1.1}$$

where *E*(0) is the measurement taken at the beginning, or time 0, and *E*(*t*) is that taken at a later time *t*.[1] However, as instruments are prone to changes with time due to degradation or ageing, their precision and sensitivity degrade, giving rise to systematic errors in the measurements. This is clearly illustrated in Fig. 1.1 which gives the fluorescence (*E*) of enamel surfaces that have been exposed to acid challenge with increasing durations and those that have not (*R*). To minimize biological variations between samples, *E* and *R* can be measurements from different regions within the same sample.

Demineralization of enamel decreases the intensity of the fluorescence detected. In theory, the fluorescence of the control or reference surfaces should remain constant; but because of the aforementioned systematic errors associated with the measurements, in practice they also change with time. In Fig. 1.1, we can clearly see a gradual reduction in the measured fluorescence of the reference surfaces with time, indicating systematic changes in the instrument or measuring conditions.

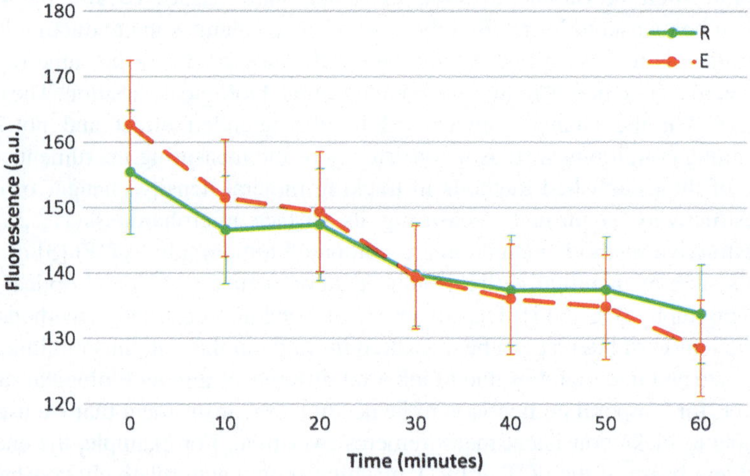

Fig. 1.1 Fluorescence (arbitrary unit) measured for enamel surfaces with (*E*) and without (*R*) demineralization challenge

[1]In mathematical language, *E*(*t*) means *E* is a function of the independent variable *t*.

This problem is usually solved by calibrating the instrument at regular intervals using standard samples. However, this may not always be practically possible. Without calibration of the instrument, the systematic errors must be estimated and corrections made for them in the measurements. As shown below, measurements from the reference samples or regions that have not been subjected to acid challenge can provide such corrections.

In passing, we would like to mention that random errors due to inherent differences among biological samples are addressed by having a sufficiently large sample size that is representative of the population. But random errors will not be discussed further here.

1.3 Corrections for Fractional Changes

To get a more reliable estimate of the fractional change in mineral density, Eq. (1.1) is modified to become

$$\frac{\Delta E}{E} = \frac{E(t) - R(t)}{R(t)}. \tag{1.2}$$

The assumption is that, at each time point, $E(t)$ and $R(t)$ suffer from similar measurement errors which can be eliminated by simple subtraction. This is not the case with $E(0)$ and $E(t)$, which were obtained at different time points and hence contain different systematic errors. Figure 1.2 compares the fractional changes in fluorescence calculated for the data in Fig. 1.1 using Eqs. (1.1) and (1.2) separately. As can be seen, because of the increasing systematic errors associated with the instrument, Eq. (1.1) overestimates the fractional changes (Fig. 1.2, Curve a).

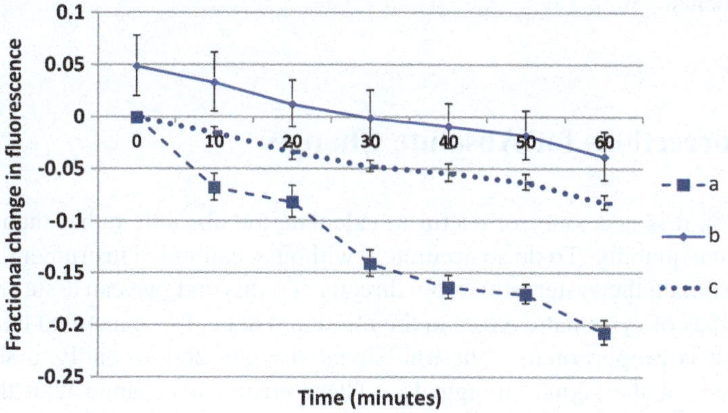

Fig. 1.2 Fractional changes in fluorescence without corrections (**a**), with correction for temporal systematic errors only (**b**), and with corrections for both temporal and spatial errors (**c**)

However, Eq. (1.2) contains errors, too, due principally to spatial differences in mineral density caused by tissue heterogeneity. This can be clearly seen from the nonzero fractional change at time 0 (Fig. 1.2, Curve b). To account for such spatial differences, Eq. (1.2) can be modified as

$$\frac{\Delta E}{E} = \frac{E(t) - R(t)\frac{E(0)}{R(0)}}{R(t)\frac{E(0)}{R(0)}} = \frac{E(t) - C_s.R(t)}{C_s.R(t)} = \frac{E(t) - R(E|t)}{R(E|t)}, \tag{1.3}$$

where $C_s = \frac{E(0)}{R(0)}$ is the correction factor for spatial errors, with $E(0)$ and $R(0)$ being measurements of the exposed and reference regions, respectively, at time 0 prior to acid challenge. Their difference is, therefore, purely due to the inherent heterogeneity of virgin tissues. $C_s.R(t) = R(E|t)$ can be considered as measurement at time t of the reference area/sample adjusted to account for spatial/biological differences in initial properties with the exposed area/sample.

Alternatively, Eq. (1.3) can be rewritten as

$$\frac{\Delta E}{E} = \frac{E(t) - R(t)\frac{E(0)}{R(0)}}{R(t)\frac{E(0)}{R(0)}} = \frac{E(t) - C_t.E(0)}{C_t.E(0)} = \frac{E(t) - E(0|t)}{E(0|t)}, \tag{1.4}$$

where $C_t = \frac{R(t)}{R(0)}$ is the correction factor for temporal errors, with $R(t)$ and $R(0)$ being measurements of the reference area at time 0 and time t, respectively. Their difference is, therefore, purely due to systematic changes in the instrument. $C_t.E(0) = E(0|t)$ can be considered as a measurement of the exposed area/sample at time 0 obtained using the instrument at t. Note that we can get Eq. (1.4) by applying C_t directly to Eq. (1.1), and $E(0|t) = R(E|t) = R(t)\frac{E(0)}{R(0)}$.

From the above examples, we can see how measurements from reference areas/samples can provide a relative rather than absolute "internal calibration" for our measurements.

1.4 Corrections for Absolute Changes

Sometimes, it is necessary or useful to calculate the absolute rather than fractional changes of a quantity. To do so accurately without a calibrated instrument, we would need to estimate the systematic errors directly. To this end, we can assume that there are two types of systematic errors in the measured data (To be justified later): firstly, a gain that is proportional to the true signal strength and secondly, a shift that is independent of the signal strength [5]. These errors can change with the time of measurement. Thus, in general, the effect of these errors on the measurements can be expressed as:

$$E(t) = k_p(t)e(t) + k_s(t) \tag{1.5}$$

where $E(t)$ and $e(t)$ are the measured and true values, respectively, at time t; $k_p(t)$ is the proportionality constant for the gain error and $k_s(t)$ is the shift error. These errors can be determined if standard samples with known true values are available. In the absence of such standard samples, we can use the following method instead. Again, measurements from reference samples will be utilized.

Applying Eq. (1.5) to measurements from a reference sample that does not change with time gives

$$R(t) = k_p(t)r + k_s(t) \tag{1.6}$$

where r is the true reference value, which is independent of time, and $R(t)$ is the measured reference value, which does change with time due to the time-dependent systematic errors.

At time 0,

$$R(0) = k_p(0)r + k_s(0). \tag{1.7}$$

Rearranging gives

$$r = \frac{R(0) - k_s(0)}{k_p(0)}. \tag{1.8}$$

Substituting Eqs. (1.8) into (1.6) gives

$$R(t) = k_p(t)\frac{R(0) - k_s(0)}{k_p(0)} + k_s(t)$$

$$= K_p(t)R(0) + K_s(t), \tag{1.9}$$

where $K_p(t) = \frac{k_p(t)}{k_p(0)}$ and $K_s(t) = k_s(t) - \frac{k_p(t)k_s(0)}{k_p(0)}$. These are the additional gain and shift introduced by the instrument at time t to the reference data, $R(0)$, obtained at time 0. Note that Eq. (1.9) no longer contains the true value, r, which cannot be established without a calibrated instrument. It only contains measurements of the reference samples collected at time 0 and time t.

Similarly, we can show that, in general

$$E(t) = K_p(t)E(t|0) + K_s(t), \tag{1.10}$$

where $E(t|0)$ is the measured value of $e(t)$, the true value of the exposed area at time t, using the instrument from time 0, i.e.,

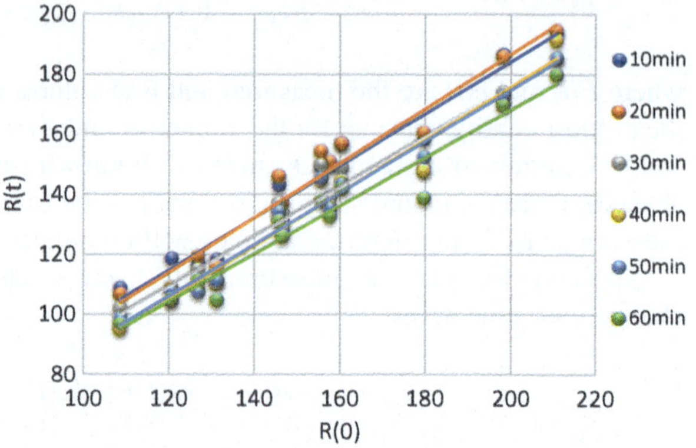

Fig. 1.3 $R(t)$ versus $R(0)$ plots of a set of reference data. The additional gain (K_p) and shift (K_s) are determined from the slope and y-intercept, respectively, of the regression line for each time point

$$E(t|0) = k_p(0)e(t) + k_s(0). \tag{1.11}$$

Rearranging Eq. (1.10) gives

$$E(t|0) = \frac{E(t) - K_s(t)}{K_p(t)} \tag{1.12}$$

This is the equation that can be used to adjust the measured absolute values to account for the time dependency of the instrument's systematic errors, provided we know the additional gain and shift. After the adjustment, all the data will be as if they were obtained using the same instrument at time 0.

The additional gain and shift, i.e., $K_p(t)$ and $K_s(t)$, can be obtained from the linear regression of the graph of $R(t)$ versus $R(0)$, i.e., measurements of the reference samples collected at time t and time 0, respectively, with $K_p(t)$ being the slope and $K_s(t)$ the y-intercept; see Eq. (1.9). Figure 1.3 shows such graphs for the reference samples in Fig. 1.1 at different time points. As can be seen, the data points can be well fitted with straight lines, thus *justifying the assumptions made in formulating* Eq. (1.5). Further, these straight lines have similar slopes, indicating that the additional gain is probably a constant. The additional shift, on the other hand, does increase with time. Figure 1.4 shows the absolute fluorescence values before and after correction using Eq. (1.12). The overestimation of the absolute changes by the systematic errors can clearly be seen.

What is the implication of the systematic errors for the fractional change? Using Eqs. (1.2), (1.5), and (1.6), the fractional change of E at time t can be expressed as

$$\frac{E(t) - R(t)}{R(t)} = \frac{[k_p(t)e(t) + k_s(t)] - [k_p(t)r + k_s(t)]}{k_p(t)r + k_s(t)} = \frac{k_p(t)[e(t) - r]}{k_p(t)r + k_s(t)}. \tag{1.13}$$

Should the shift (k_s) be much smaller than the gain (k_p),

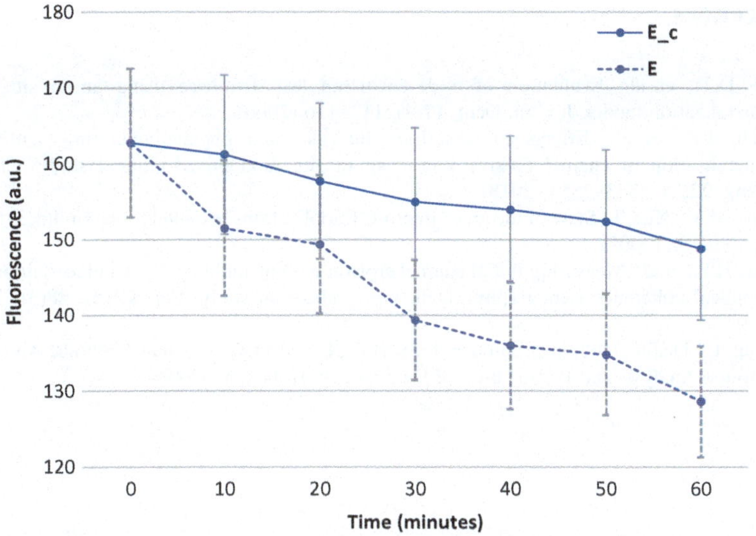

Fig. 1.4 Absolute fluorescence values of enamel surfaces before (E) and after (E_c) correction for gain and shift errors

$$\frac{E(t) - R(t)}{R(t)} \approx \frac{e(t) - r}{r}. \tag{1.14}$$

As can be seen from Eq. (1.14), the estimated fractional change, in this case, is not very sensitive to the systematic errors of the instrument. Therefore, if only fractional changes are required, it would be advantageous to increase the gain of an instrument. However, bearing in mind that the true fractional change is $\frac{e(t)-e(0)}{e(0)}$, the estimation will still be sensitive to the spatial error, i.e., how different the reference data r is from $e(0)$, the true value of the exposed area at time 0.

1.5 Concluding Remarks

In this chapter, we have illustrated through a simple example the usefulness of a mathematical model, i.e., an explicit mathematical equation that relates the parameters of interest. Since it is quantitative, it enables prediction to be made of the dependent parameter for conditions that are different from those of the initial experiment. Very often, this could lead to an improvement of the task at hand or new ideas for investigation, as we shall see in the following chapters.

References

1. Zero, D.T., et al.: Evaluation of a desensitizing test dentifrice using an in situ erosion remineralization model. J. Clin. Dent. **17**(4), 112–116 (2006)
2. Creeth, J.E., et al.: Effects of a sodium fluoride- and phytate-containing dentifrice on remineralisation of enamel erosive lesions-an in situ randomised clinical study. Clin. Oral Investig. **22**(7), 2543–2552 (2018)
3. Swain, M.V., Xue, J.: State of the art of micro-CT applications in dental research. Int. J. Oral Sci. **1**(4), 177–188 (2009)
4. Chew, H.P., et al.: Measuring initial enamel erosion with quantitative light-induced fluorescence and optical coherence tomography: an in vitro validation study. Caries Res. **48**(3), 254–262 (2014)
5. Kirkup, L.: Dealing with uncertainties. In: Kirkup, L. (ed.) Experimental Methods. An Introduction to the Analysis and Presentation of Data. Wiley, Hoboken (1994)

Chapter 2
Endodontic Treatment and Vertical Root Fracture

2.1 Introduction

When the dental pulp tissues become severely infected, endodontic treatment is deemed necessary. The procedure uses a tapered, bendable file with screw threads to access the root canal in preparation for the removal of infected tissues. The root canal, especially at the orifice, will be enlarged and reshaped in the process to ease debris removal of the apical portion of the canal. However, during root canal access and cleaning, or the subsequent compaction of root-filling materials to seal off the root, excess force could be applied which causes the root to fracture vertically [1]. A tooth with this so-called vertical root fracture, if it is long, could be difficult to restore, and extraction is normally indicated.

In order to determine the maximum force that can be used to clean and reshape root canals (especially with rotary instruments) or to seal with compaction materials, fracture tests have been carried out with whole teeth or thin sections of tooth roots [2]. As is common in dental materials research, the experiments explored, through statistical analysis of the results, the correlations between the fracture load and the geometrical factors that were thought to be important. These included the root width, canal width, canal taper, etc. [2].

Counterintuitively, perhaps, the fracture load was found to increase, or at least not to decrease, with increasing canal width [2]. Figure 2.1 shows the fracture loads of thin root sections fractured by a tapered rod inserted through the circular hole formed by root canal enlargement. As can be seen, the fracture load tends to increase with the enlarged root canal diameter, even though the linear relationship is not strong.

In order to understand this unexpected result, we will take a more mechanistic approach. We will make use of the most fundamental principle of mechanics, i.e., equilibrium or the balance of forces, to analyse the above problem. This requires forces to be resolved into their components in particular directions. We will also introduce a simple model for friction, which occurs where the tapered rod is in contact with the root section. The result is a mathematical model or, more precisely,

© Springer Nature Switzerland AG 2020

A. Fok, H. P. Chew, *Mathematical Models for Dental Materials Research*,
https://doi.org/10.1007/978-3-030-37849-3_2

a mechanistic equation that relates the maximum allowable load to the key geometrical and material properties of the specimens. An attempt will then be made to validate the equation by comparing its predictions for the fracture load with the experimental data shown in Fig. 2.1.

2.2 Theory

Figure 2.2 depicts a pictorial model for the experiment that produced the results shown in Fig. 2.1. A root dentin disc is loaded by a circular metal rod inserted through the enlarged circular root canal of diameter *D*. The metal rod and the

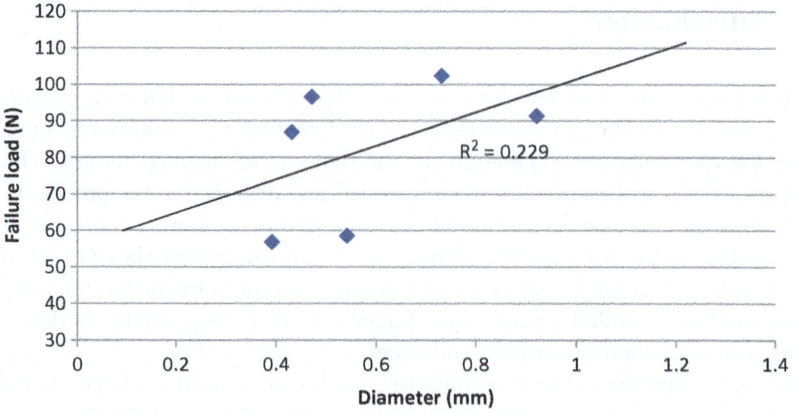

Fig. 2.1 Fracture load as a function of enlarged root canal diameter for root sections loaded by a tapered rod [3]

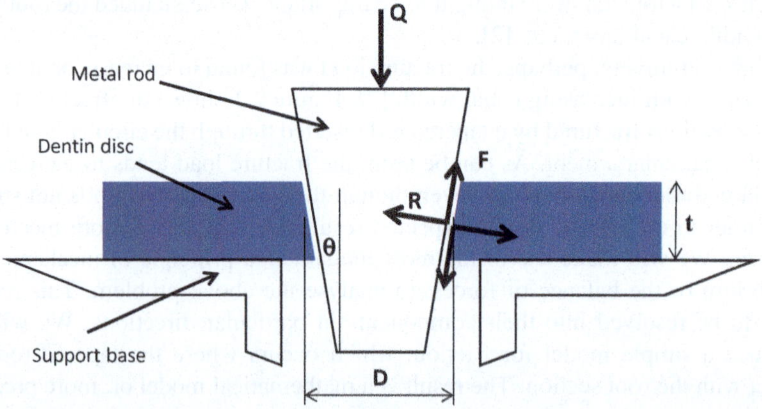

Fig. 2.2 Schematic diagram of dentin disc loaded by a metal rod inserted through the central root canal

enlarged root canal have the same taper angle, θ (radians). It is useful to recognize that the disc is being expanded by the rod while the rod is being compressed by the disc at the same time. Newton's Third Law states that to every action there is an equal but opposite reaction. Thus, at the surfaces of contact between the rod and the disc, there is a pair of equal and opposite forces that are normal to the surfaces (R) and a pair of equal and opposite frictional forces that are parallel to the surfaces (F). These internal forces balance the external applied load Q and the reaction force from the support base to keep the rod and disc, respectively, in static equilibrium.

Consider the balance of forces in both the vertical and horizontal directions for either the rod or the root dentin disc shown in Fig. 2.2. To do so, we need to visualize the object together with all the external and internal forces that act on it in isolation. This mental exercise takes some practice but it is an essential and very powerful tool in mechanics.

Let R and F be the direct and shear stresses, or force intensity (force per unit area[1]), that are, respectively, normal and parallel to the contact surfaces. The latter have a common area of $\frac{\pi D t}{\cos\theta}$, where t is the disc thickness. Vertically for the rod, the vertical components of the internal forces (stress $\times$ area) will balance out the external load Q, i.e.,

$$Q = (F\cos\theta + R\sin\theta)\frac{\pi D t}{\cos\theta}. \tag{2.1}$$

If we want to consider the equilibrium of the disc instead, we would need to replace the support base with a reaction force that is equal and opposite to the external load Q. But we will arrive at the same equation (2.1).

Horizontally, as they act around the entire circumference symmetrically, the horizontal components of the interfacial stresses balance themselves out: for each horizontal force that acts at a point on the surface of the rod, there is an equal but opposite force on the other side of the circumference.

To proceed with the analysis, we need to determine the pressure, p (also force per unit area), exerted by the metal rod on the inner surface of the disc, or vice versa. p can be found from the horizontal components of the interfacial stresses, i.e.,

$$p = -F\sin\theta + R\cos\theta. \tag{2.2}$$

We have two equations (2.1) and (2.2), but three unknowns (p, F, and R). In order to solve our problem, we need to find another equation.

The frictional force is assumed to be proportional to the reaction force, i.e.,

$$F = \mu R, \tag{2.3}$$

where μ is the frictional coefficient. This is a very simple model that again makes use of the straight-line equation, $y = mx + c$, but with the added simplification that there

[1]We are assuming that the forces are evenly distributed over the surfaces in contact.

is no "y-intercept." μ is assumed to be dependent on the nature of the surfaces in contact but independent of the forces and the contact area.

Using Eq. (2.3), Eqs. (2.1) and (2.2) can then be rewritten as follows to eliminate F:

$$Q = R(\mu \cos\theta + \sin\theta) \times \frac{\pi Dt}{\cos\theta}, \tag{2.4}$$

$$p = R(-\mu \sin\theta + \cos\theta). \tag{2.5}$$

R can then be eliminated also by dividing Eq. (2.4) by (2.5), which gives

$$\frac{Q}{p} = \left(\frac{\pi Dt}{\cos\ \theta}\right) \frac{\mu \cos\theta + \sin\theta}{-\mu \sin\theta + \cos\theta},$$

or

$$Q = \left(\frac{\pi Dtp}{\cos\theta}\right) \frac{\mu + \tan\theta}{1 - \mu \tan\theta}. \tag{2.6}$$

The internal pressure, p, will induce tensile stresses in the circumferential direction and compressive stresses in the radial direction within the expanded root dentin disc, but it is the tensile stresses that cause fracture. According to [2], the maximum circumferential stress in the disc occurs at the surface in contact with the rod and is related to the internal pressure, p, via

$$\sigma_{\max} = kp = \frac{kQ \cos\ \theta}{\pi Dt} \left(\frac{1 - \mu \tan\theta}{\mu + \tan\theta}\right), \tag{2.7}$$

where k is a stress concentration factor that depends on the ratio between the root canal diameter and the planar dimensions of the root section. As expected, the higher is the external load Q, the higher the stress. Hence, at fracture, i.e., when $\sigma_{\max}$ is equal to the fracture strength of dentin (σ_f), the vertical force reaches its critical value of

$$Q_f = \frac{\pi Dt}{\cos\theta} \left(\frac{\sigma_f}{k}\right) \frac{\mu + \tan\theta}{1 - \mu \tan\theta}. \tag{2.8}$$

When the root canal diameter is much smaller than the planar dimensions of the root section, $k \approx 1$ [4]. Hence,

$$Q_f \approx \left(\frac{\pi Dt\sigma_f}{\cos\theta}\right) \frac{\mu + \tan\theta}{1 - \mu \tan\theta}. \tag{2.9}$$

Further, when the taper is small, Eq. (2.9) simplifies to:

$$Q_f \approx \pi D t \sigma_f \left(\frac{\mu + \theta}{1 - \mu\theta} \right) \approx \pi D t \sigma_f (\mu + \theta). \tag{2.10}$$

Numerical simulation of the test using the finite element method [3] has produced results that follow closely Eq. (2.9), thus justifying the various assumptions made in its derivation.

2.3 Validation

Equation (2.9) shows that the fracture load of a root section is proportional to its root canal diameter. It also shows that the fracture load increases with the taper. Both of these theoretical predictions agree with the findings by Holcomb et al. [2]. Mechanically, the larger the canal diameter, the larger the circumferential area there is to distribute the contact load, and hence the lower the pressure and stresses in the dentin.

The experimental data in Fig. 2.1 do not quite fall on a straight line as Eq. (2.9) predicts. This is because the specimens came from different parts of the roots: those with a smaller canal diameter came from the bottom of the root (apical region) and those with a bigger canal diameter came from the top (coronal region). As a result, they had different fracture strengths, with apical dentin being roughly twice as strong as coronal dentin [5]. When these, together with the different tapers, were taken into account, Eq. (2.9) correctly predicts the trends of the data, even though there appears to be a systematic error in the predictions (Fig. 2.3).

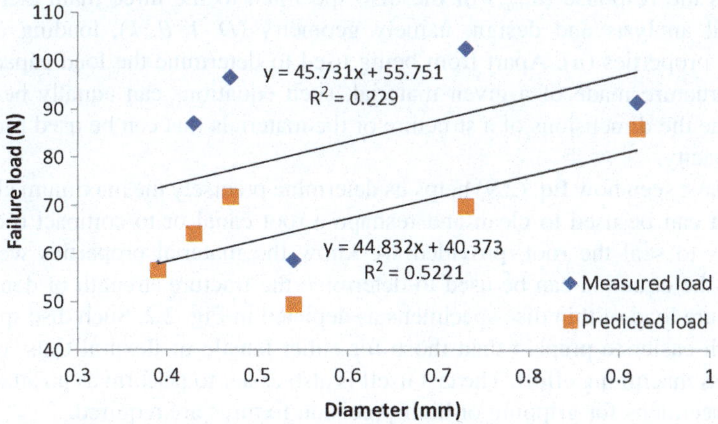

Fig. 2.3 Experimental and predicted failure loads versus root canal diameter [3]

The fracture strengths used in the predictions are based on data from tensile tests [5], which tend to give lower values than flexural tests, for example. It is typical of brittle materials such as dentin that their fracture strength is dependent on the test configuration or, more precisely, the stress gradient that is present in the specimen. The more nonuniform the stress distribution, the higher the stress required to cause fracture. Indeed, the root dentin disc has a nonuniform stress distribution [3], with the stresses reducing with the radial distance from the centre. It is therefore not surprising that the predictions, which are based on the tensile strength, underestimated the fracture loads.

2.4 Concluding Remarks

Described in this chapter is a simple example that clearly demonstrates the power of mathematical modelling. The process is not dissimilar to the other forms of hypothesis-driven research; except that in mathematical modelling the formation of a hypothesis is by means of an explicit mathematical equation that is based on suitable laws of nature. The model helps us understand mechanistically our experimental results, some of which may seem counterintuitive in the first place. It can also provide us with guidelines for clinical practice and for the design and construction of surgical instruments. Moreover, the model may usefully lead to new experiments, such as the disc fracture test analysed in this chapter, which can readily provide us with essential mechanical properties of the materials under investigation. However, it must be stressed that all models have limitations. In this case, the model will break down when the canal width is no longer small compared to the outer dimensions of the root section. The dimensional ratio beyond which the model is no longer applicable needs to be determined.

Equation (2.7) is a typical equation often encountered in engineering mechanics. It relates the response ($\sigma_{\max}$) of the disc specimen to the three main elements of structural analysis and design, namely geometry (D, t, θ, k), loading (Q), and material properties (μ). Apart from being used to determine the load capacity of a given structure made of a given material, such equations can equally be used to determine the dimensions of a structure or the materials that can be used for a given load capacity.

We have seen how Eq. (2.9) helps us determine precisely the maximum allowable load that can be used to clean and reshape a root canal or to compact root-filling materials to seal the root, provided we know the material properties well. Conversely, the equation can be used to determine the fracture strength of dentin from the fracture load of thin disc specimens as depicted in Fig. 2.2. Such disc specimens are much easier to prepare than those for either tensile or flexural tests, requiring much less machining effort. The test itself is also easier to perform as no attachments to the specimens for gripping or high-precision fixtures are required.

Incidentally, the loading induced by a dental implant in the surrounding bone is not too dissimilar to that induced by the metal rod in the dentin disc. A similar model can, therefore, be constructed to analyse the former problem.

References

1. Haueisen, H., et al.: Vertical root fracture: prevalence, etiology, and diagnosis. Quintessence Int. **44**(7), 467–474 (2013)
2. Holcomb, J.Q., Pitts, D.L., Nicholls, J.I.: Further investigation of spreader loads required to cause vertical root fracture during lateral condensation. J. Endod. **13**(6), 277–284 (1987)
3. Munari, L.S., Bowles, W.R., Fok, A.S.L.: Relationship between canal enlargement and fracture load of root dentin sections. Dent. Mater. **35**(5), 818–824 (2019)
4. Pilkey, W.D., Pilkey, D.F.: Peterson's Stress Concentration Factors Handbook, p. Chart 4.2. Wiley, Hoboken (2008)
5. Mannocci, F., et al.: Density of dentinal tubules affects the tensile strength of root dentin. Dent. Mater. **20**(3), 293–296 (2004)

Chapter 3
Debonding of Resin Composite Restorations

3.1 Introduction

Modern dental restorations used for replacing damaged or carious tooth tissues are often made of light-cured resin composites that can be bonded to the remaining tooth structure. When cured, such materials will shrink due to polymerization condensation. It is recognized that the resulting shrinkage stress could cause debonding of the restoration [1, 2], which would lead to its early failure [3] through the loss of structural integrity or development of recurrent caries [4]. Therefore, there has been much interest and work done in finding out how a restoration will shrink within the tooth cavity preparation, the magnitude of the shrinkage stresses generated, and the conditions under which debonding will occur.

A common approach taken by medically or biologically trained researchers in analysing shrinkage stress is to hypothesize that shrinkage or its consequences are related to variables such as material properties, cavity designs, curing techniques, etc. A series of experiments would then be carried out and statistical analysis performed to determine whether the hypothesis should be rejected or accepted [5]. Depending on the outcome of this statistical analysis, further analysis may be performed to obtain correlation between the variables.

Such an empirical approach is useful and necessary in the beginning of any investigation when the relationships between the different variables are unclear due to the lack of a mechanistic understanding—we have seen such an example in Chap. 2 for analysing fracture of tooth roots. However, when one has gathered enough information from experiments to have a relatively good idea of what the key variables are and their approximate mathematical relationships, a more analytical approach would be desirable for further investigation. As illustrated in the previous chapter, the resulting analytical expression of the key variables can provide new insights to the problem, make quantitative predictions for conditions beyond those that generate the original data or suggest new ways of doing things.

© Springer Nature Switzerland AG 2020

A. Fok, H. P. Chew, *Mathematical Models for Dental Materials Research*,
https://doi.org/10.1007/978-3-030-37849-3_3

In this chapter, we will make use of the straight-line equation, again, to analyse shrinkage displacements in resin composite restorations. We will see how debonding at the floor of a cavity can change the shrinkage displacement profile of the restoration, a result that we can exploit for the detection of debonding. We will also introduce, for the first time, differential equations, which are essential for analysing problems with variables that change with space and/or time. Since simple analytical solutions cannot always be found for differential equations, we will show how to solve them approximately using the concept of slope. The same technique will be used over and over again in the remaining chapters of this book to obtain approximate solutions for problems with increasing complexity. The reader is therefore advised to spend some time to get themselves acquainted with the materials in this chapter before moving on to the later parts of the book. In particular, they should try to reproduce the numerical solution by implementing the algorithm described in a spreadsheet software, such as Microsoft Excel.

3.2 Theory

Experiments have indicated that the total amount of shrinkage (δ), or reduction in length, of a specimen is proportional to its original length (l). We can, therefore, use the following simple mathematical model to relate these two variables:

$$\delta = \varepsilon_s l, \tag{3.1}$$

where ε_s, the dimensionless constant of proportionality, is known as the shrinkage strain. It can be seen from rearranging Eq. (3.1) that the definition of ε_s is shrinkage per unit length.

Now, how a resin composite specimen is constrained will determine how it will deform when it shrinks. For example, when a strip of resin composite of length l is fixed at its mid-point, its two free ends will move towards its centre with equal but opposite displacement of $\frac{\varepsilon_s l}{2}$. On the other hand, when one end of the specimen is fixed, the other end will move towards that fixed end by an amount of $\varepsilon_s l$. The constraints are known, in mathematical terms, as boundary conditions. Therefore, how a structure will deform will depend on its boundary conditions.

In general, if we consider a point that is x units away from a reference point, under shrinkage, the two will move towards each other by an amount of $\varepsilon_s x$. This is relative displacement. Mathematically, we can write the absolute displacement of any point due to shrinkage as a linear function of its position x, namely,

$$\delta(x) = -\varepsilon_s x + C. \tag{3.2}$$

where C is a constant that depends on the boundary conditions. The minus sign indicates that a positive shrinkage reduces the distance between two points. Eq. (3.2) is, of course, the straight-line equation, with $-\varepsilon_s$ being the slope and C the y-intercept.

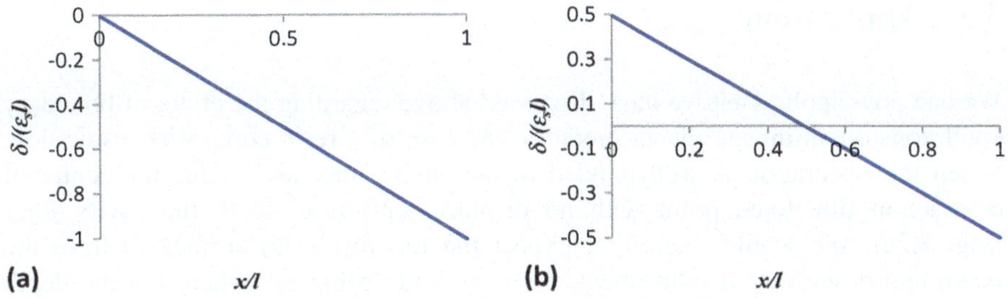

Fig. 3.1 Shrinkage displacement profiles of a resin composite strip fixed at (**a**) one end and (**b**) the centre

It can be seen that $\delta(0) = C$, i.e., C is the absolute displacement of the reference point or origin.

If we assume that the left-hand end of our specimen is the origin (where $x = 0$) and is fixed (Fig. 3.1a), then $\delta(0) = C = 0$ and

$$\delta(x) = -\varepsilon_S x. \tag{3.3}$$

As expected, the right-hand end where $x = l$ will displace by $\delta(l) = -\varepsilon_s l$, i.e., it will move to the left towards the origin with a displacement of magnitude $\varepsilon_s l$.

If, on the other hand, the midpoint (where $x = l/2$) of our specimen is fixed (Fig. 3.1b), then $\delta(l/2) = -\varepsilon_s \left(\frac{l}{2}\right) + C = 0$ or $C = \frac{\varepsilon_s l}{2}$, and

$$\delta(x) = -\varepsilon_S \left(x - \frac{l}{2} \right). \tag{3.4}$$

Now, the origin will move with a positive displacement $\delta(0) = \frac{\varepsilon_s l}{2}$, i.e., to the right, while the right-hand end will move with a negative displacement of the same magnitude, i.e., to the left.

Figure 3.1 shows graphically the shrinkage displacement profiles for these two cases. To save us from writing down units, the distance x is normalized (divided) by the length, l, of the specimen while the shrinkage displacement is normalized by its maximum value, $\varepsilon_s l$. Note that the two graphs have the same slope because of the same shrinkage strain but different y-intercepts because of the different boundary conditions. In the specimen where the left-hand end is fixed, all the points have negative displacements. In the specimen where the midpoint is fixed, the left half has positive displacements while the right half has negative displacements. In both cases, the further away a point is from the fixed point, the greater the displacement.

3.3 Application

We can now apply what we have discussed above regarding the effect of boundary conditions on shrinkage displacement to the case of a resin composite restoration. When the restoration is well bonded to the cavity floor and walls, the centre of contraction (the fixed point with no displacement) is close to the cavity floor (Fig. 3.2a). We would, therefore, expect the top (occlusal) surface to have the maximum downward displacement, the magnitude being $\varepsilon_s l$, where l is the depth of the restoration (Fig. 3.1a). However, when there is debonding at the cavity floor, the centre of contraction will shift to the geometrical centre of the restoration (Fig. 3.2b). We then expect the displacement of the occlusal surface to have a magnitude of $\frac{\varepsilon_s l}{2}$ only, with the other half of the total shrinkage being taken up by the upward movement at the cavity floor (Fig. 3.1b).

This raises the possibility of using the surface displacement of dental restoration to assess its interfacial integrity: if the measured value is roughly half that expected from the total shrinkage, then debonding at the cavity floor has probably occurred.

Figure 3.2 shows the shrinkage displacement vectors obtained via digital image correlation (DIC) for model restorations with or without artificially induced "pulpal floor debonding" [6]. The shift in the centre of contraction and the symmetry in the displacement profile between the top and bottom of the restoration following debonding at the floor can clearly be seen.

Figure 3.3 compares the theoretical and experimental shrinkage displacements as a function of position for both the bonded and debonded model restorations in Fig. 3.2. The theoretical predictions are based on Eqs. (3.3) and (3.4). As can be seen, there is good agreement between theory and experiment. The difference between the two is likely due to the compliance or partial debonding of the adhesive

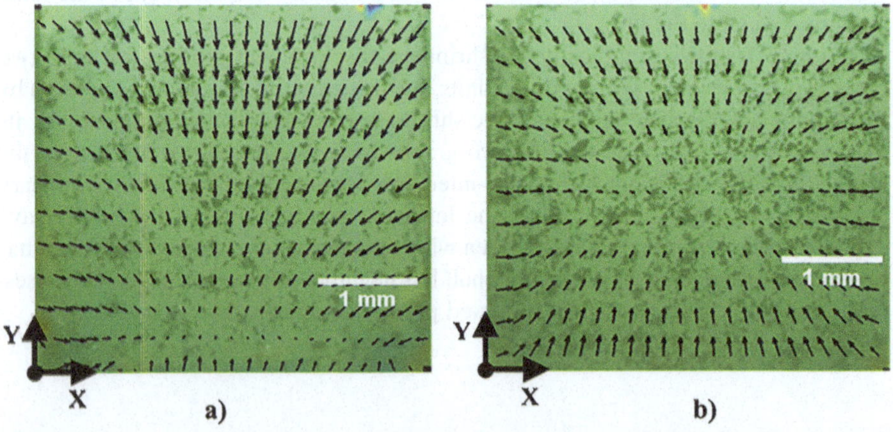

Fig. 3.2 DIC-calculated shrinkage displacement vectors (10× magnification) for (**a**) fully bonded restoration and (**b**) restoration debonded at the floor [6]

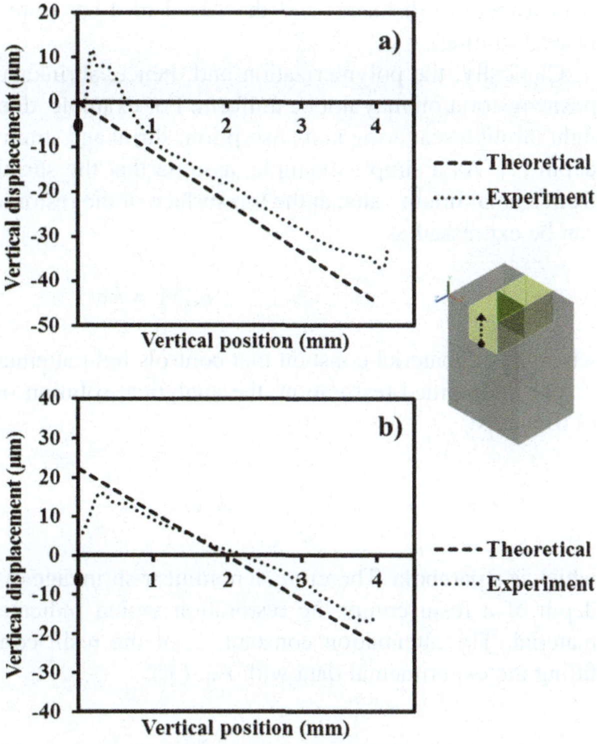

Fig. 3.3 Comparison between experimental and analytical displacements along the vertical centreline for (**a**) bonded and (**b**) debonded restorations from bottom (0 mm) to top (4 mm) [6]

layer used to bond the resin composite to the cavity surface and possible slumping of the material under gravity.

3.4 Nonuniform Shrinkage

The mathematically minded reader will recognize that Eq. (3.2) is the general solution for the first-order differential equation

$$\frac{\mathrm{d}\delta}{\mathrm{d}x} = -\varepsilon_{\mathrm{s}}. \tag{3.5}$$

One can prove this by differentiating Eq. (3.2) with respect to x. In fact, Eq. (3.5) is the more general definition of shrinkage strain, which is the rate of change of displacement due to shrinkage with respect to distance. In general, a differential equation describes the rate of change of a variable with respect to an independent variable such as time or distance. The rate of change may be a constant, a function of the independent variable, a function of the dependent variable itself, or a combination of the above. Daily examples include the slope of a hill (rate of change of height with respect to distance) and the speed of a car (rate of change of distance with respect to time).

Clinically, the polymerization and, hence, shrinkage strain within a resin composite restoration may not be uniform. For example, due to the attenuation of curing light through scattering and absorption, shrinkage strain will reduce with restoration depth [7]. As a simple example, assume that the shrinkage strain reduces linearly from its maximum value at the top surface of the restoration to zero at the floor. This can be expressed as

$$\varepsilon_s(x) = kx, \tag{3.6}$$

where k is a material constant that controls light attenuation.

For the bonded restoration, the analytical solution of Eq. (3.5) with assumption (3.6) yields

$$\delta(x) = -k\frac{x^2}{2}, \tag{3.7}$$

which is a parabola. Therefore, a nonlinear shrinkage displacement profile along the depth of a resin composite restoration would indicate nonuniform curing of the material. The attenuation constant, k, of the resin composite can be obtained by fitting the experimental data with Eq. (3.7).

3.5 Numerical Solution

Very often, it is not possible to obtain a simple analytical solution, such as Eq. (3.7), for a differential equation that has more complicated terms. In such cases, one could make use of numerical methods to obtain approximate solutions which, when calibrated with experimental data, can yield estimates for the material constants, for example. The principle on which the numerical methods are based is rather simple.

We are all familiar with the concept of slope, which is the change in height over a lateral distance. So the change in height, as we climb up a mountain, say, can be calculated by multiplying the slope with the lateral distance covered, assuming these are the only information available. If the slope changes along the way, we will need to perform our calculations regularly and sum up the results to get the cumulative change in height, i.e., our altitude. The faster the slope changes, the more frequent we would need to calculate in order to ensure accuracy. For our resin composite restoration, the shrinkage displacement is the cumulative height, the shrinkage strain is the slope and the axial position along the restoration is the lateral distance.

For illustration, Eq. (3.5) with assumption (3.6) can be approximated as

$$\frac{\Delta\delta}{\Delta x} = -kx, \tag{3.8}$$

where $\Delta\delta$ and Δx are small but finite changes in shrinkage displacement and depth, respectively, instead of the infinitely small changes in the differential equation. The change in shrinkage displacement between two points with distance Δx is thus

$$\Delta\delta = -kx\Delta x. \tag{3.9}$$

If the restoration is divided into a number of levels, Eq. (3.9) will allow us to calculate successively the displacement at each of these levels provided we know the value at one of them, which will be our starting point or boundary condition.

Again, using the bonded restoration as an example, if we divide it into four levels, we can construct the following table to estimate the displacement profile starting from the floor of the restoration where displacement is zero:

x	$\Delta\delta$	δ
x_0		$\delta_0 = 0$
x_1	$\Delta\delta_1 = -kx_1(x_1 - x_0)$	$\delta_1 = \delta_0 + \Delta\delta_1$
x_2	$\Delta\delta_2 = -kx_2(x_2 - x_1)$	$\delta_2 = \delta_1 + \Delta\delta_2$
x_3	$\Delta\delta_3 = -kx_3(x_3 - x_2)$	$\delta_3 = \delta_2 + \Delta\delta_3$
x_4	$\Delta\delta_4 = -kx_4(x_4 - x_3)$	$\delta_4 = \delta_3 + \Delta\delta_4$

The first column gives the positions of the top and bottom of each level; the second column gives the change in shrinkage displacement using the current shrinkage strain; and the third column gives the cumulative shrinkage displacement. Note

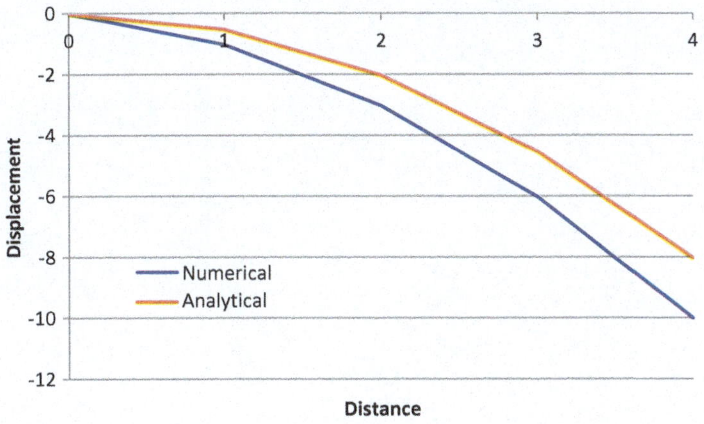

Fig. 3.4 Comparison between numerical and analytical displacements along the vertical centreline for a bonded restoration with nonuniform (linearly varying) shrinkage strain

that the shrinkage strain at the top of each level is used in the calculation, but one can use the value at the bottom of each level as well. The above table can easily be implemented with a spreadsheet. The result is given in Fig. 3.4 where the numerical solution with four levels is compared with the analytical solution for $k = 1$ and $l = 4$. As can be seen, the numerical solution contains errors, which will impact on the accuracy of the material constant we want to determine. We will see another example in the next chapter. The numerical errors can usually be reduced by increasing the number of levels in the calculation. However, it would be more efficient, in this case, to use the average shrinkage strain or slope within each level. The overall profile of the shrinkage displacement is, nevertheless, well captured in Fig. 3.4, despite the small number of levels used in this example.

References

1. Ferracane, J.L., Hilton, T.J.: Polymerization stress—is it clinically meaningful? Dent. Mater. **32** (1), 1–10 (2016)
2. Yamamoto, T., et al.: Contraction stresses in dental composites adjacent to and at the bonded interface as measured by crack analysis. Acta Biomater. **7**(1), 417–423 (2011)
3. Kopperud, S.E., et al.: Longevity of posterior dental restorations and reasons for failure. Eur. J. Oral Sci. **120**(6), 539–548 (2012)
4. Bernardo, M., et al.: Survival and reasons for failure of amalgam versus composite posterior restorations placed in a randomized clinical trial. J. Am. Dent. Assoc. **138**(6), 775–783 (2007)
5. Chang, M.: What constitutes science and scientific evidence: roles of null hypothesis testing. Educ. Psychol. Meas. **77**(3), 475–488 (2017)
6. Novaes Jr., J.B., et al.: Can pulpal floor debonding be detected from occlusal surface displacement in composite restorations? Dent. Mater. **34**(1), 161–169 (2018)
7. Li, J., et al.: Measurement of the full-field polymerization shrinkage and depth of cure of dental composites using digital image correlation. Dent. Mater. **25**(5), 582–588 (2009)

Chapter 4
Attenuation of Curing Light Through Resin Composite Restorations

4.1 Introduction

It is important to fully cure a restoration made of resin composite. Otherwise, the mechanical strength of the restoration will be compromised, resulting in its early failure. Also, leaching of unreacted monomers could constitute a health concern [1].

Modern dental resin composites use visible blue light for curing. As the light passes through the material, it gets absorbed and scattered. Consequently, the material at the bottom of restoration does not cure as fast as that at the top. To fully cure a resin composite restoration, therefore, the curing light must be applied for an adequate length of time. However, some cavity preparations are so deep that the light reaching the bottom of the restorations may be negligible. As a result, they need to be filled in increments [2]. The maximum thickness of resin composite that can be placed and cured at any one time in a cavity is material dependent. Nowadays, there are so-called bulk-filling resin composites that allow thicker increments to be placed and cured [3, 4].

In order to determine the maximum thickness of resin composite that can be placed and cured in a cavity at any one time, and the time required to fully cure the material, it is essential to know how quickly light diminishes (or attenuates) as it passes through the material. Roughly speaking, a resin needs to absorb a certain amount of energy to cure adequately, and the time required for curing is inversely proportional to the light intensity (power per unit area).

In this chapter, we will derive the Beer–Lambert Law of light attenuation [5] to help us answer the above practical questions. We will see from this example how one can use existing experimental data to come up with a suitable mathematical or differential equation to describe the changes of a variable with an independent variable in general; in this case, the change of light intensity with depth in translucent materials. We will also see how ensuring that the equation satisfies the boundary conditions can guide us in selecting the right form for the equation. Again, we will see how numerical methods, which can readily be implemented with MS Excel, can help us obtain approximate solutions for the differential equation.

© Springer Nature Switzerland AG 2020

A. Fok, H. P. Chew, *Mathematical Models for Dental Materials Research*,
https://doi.org/10.1007/978-3-030-37849-3_4

4.2 Theory

Figure 4.1 shows the light irradiance (or intensity), I, emerging from blocks of a resin composite of different thicknesses, x [6]. It can be considered as the reduction in light irradiance with depth due to absorption and scattering as the light passes through a single thick block of resin composite. One could be tempted to fit the data with a polynomial or exponential function of x. But this alone does not provide much understanding of the material behaviour. Nor does it allow predictions to be made for situations that are different from those used for gathering the original data.

We can see from Fig. 4.1 that the attenuation is not linear, i.e., unlike the shrinkage displacement profile with a constant shrinkage strain in Fig. 3.1, the slope, $\frac{dI}{dx}$, of the light intensity profile is not a constant. Rather, the slope is a function of x or I. Specifically, we note that the magnitude of the slope, which is always negative, reduces as x increases or as I reduces.

In formulating differential equations, one should always try to keep things simple. After all, there is no point using complicated functions when simpler ones will do. We can, therefore, try writing

$$\frac{dI}{dx} = -\frac{k}{x}, \tag{4.1}$$

or

$$\frac{dI}{dx} = -kI, \tag{4.2}$$

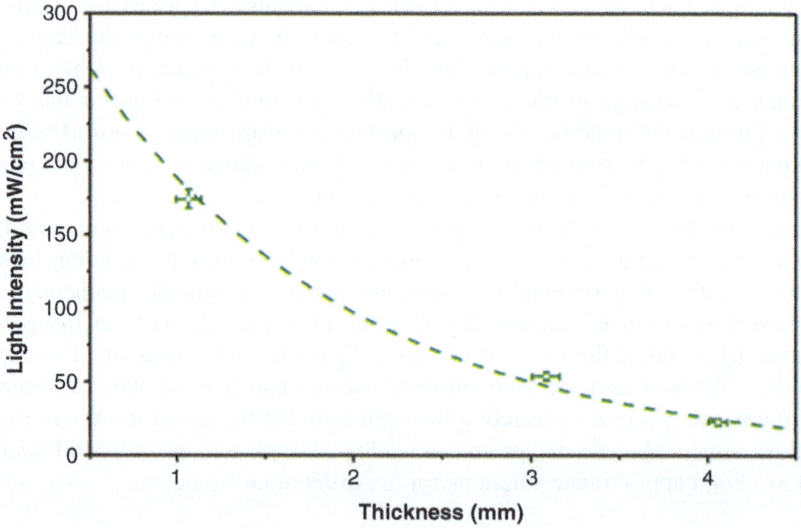

Fig. 4.1 Irradiance emitted from the end of composite blocks with different thicknesses [6]

where k is a constant. Both equations satisfy the boundary condition of the slope approaching zero when x is large or I is low. However, Eq. (4.1) will not work as it gives an infinite slope and hence an infinite irradiance at $x = 0$. This is clearly not the case because when $x = 0$, the light irradiance should equal the output of the curing light. Equation (4.2), on the other hand, will not suffer from this shortcoming if it can provide a finite solution for I at $x = 0$. We will, therefore, proceed with Eq. (4.2). The reader is encouraged to try out different forms for the differential equation and see if they can also satisfy the boundary conditions.

By dividing both sides of Eq. (4.2) by I, we can see that the underlying assumption is that the fractional reduction in I, i.e. dI/I, per unit thickness of material is a constant. And the greater is the value of k, the greater the reduction or attenuation of light intensity by the material. Should this assumption be true, we will gain a fundamental understanding of the optical behaviour of resin composites, which cannot be provided by curve fitting alone. Other forms of the differential equation may point to different material behaviours.

4.3 Numerical Solution

Before we solve Eq. (4.2) analytically, we will demonstrate again how we can use its approximation $\frac{\Delta I}{\Delta x} = -kI$ to compute numerical solutions. In the same manner as that described in Chap. 3, we rearrange the approximation of Eq. (4.2) as:

$$\Delta I = -kI\Delta x \tag{4.3}$$

to get the change in irradiance for a small but finite change in depth. Thus, if I is the irradiance at depth x, then at a deeper location $x + \Delta x$, the irradiance will become

$$I + \Delta I = I - kI\Delta x = I(1 - k\Delta x). \tag{4.4}$$

So, provided we know the incident irradiance, I_0, at $x = 0$, the irradiance at successive depths $x_1, x_2, x_3, \ldots$ can be computed as follows:

$$\begin{aligned} I_1 &= I_0 - kI_0(x_1 - x_0) \\ I_2 &= I_1 - kI_1(x_2 - x_1) \\ I_3 &= I_2 - kI_2(x_3 - x_2) \\ &\vdots \\ I_n &= I_{n-1} - kI_{n-1}(x_n - x_{n-1}), \end{aligned} \tag{4.5}$$

where n is the number of divisions the restoration is separated into for numerical computation.

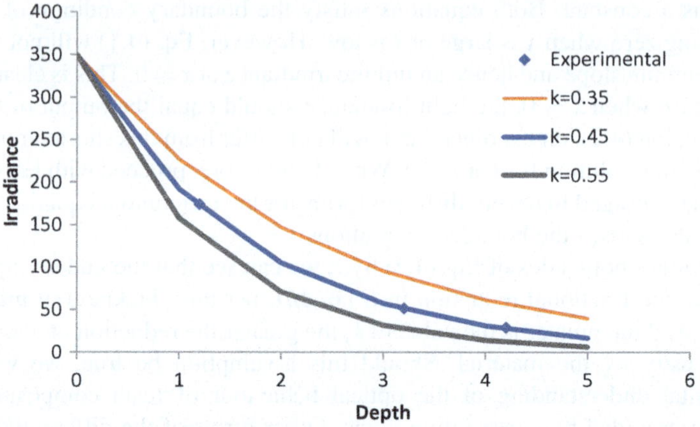

Fig. 4.2 Computed irradiance as a function of depth for $n = 5$ and different values of k

We can do this for different k values and see which one provides the best fit to the experimental data. Again, the computation can be easily implemented using MS Excel, the results of which are shown in Fig. 4.2 for $n = 5$ and three different values of k. As can be seen, the numerical solution best matches the experimental results when $k = 0.45$ for this value of n.

However, because of the finite step size used in the computation (refer to Sect. 3.4), these numerical solutions contain errors; so does k. The numerical errors can be reduced by increasing the number of divisions, which will produce a different value for k that would best fit the data. Thus, we can obtain a series of best-fit k values with increasing n and the converged k value will be our best estimate for the attenuation coefficient. Figure 4.3 shows the convergence of k with increasing n, or decreasing $1/n$. As can be seen, k has a converged value of approximately 0.58.

4.4 Analytical Solution

Equation (4.2) has the following general analytical solution:

$$\log_e I = -kx + C, \tag{4.6}$$

where C is a constant that depends on the incident irradiance. To find C, we use the boundary condition of $I = I_0$ at $x = 0$. This gives $C = \log_e I_0$, and the solution specific to our problem is:

$$\log_e I = -kx + \log_e I_0,$$

or, upon rearranging,

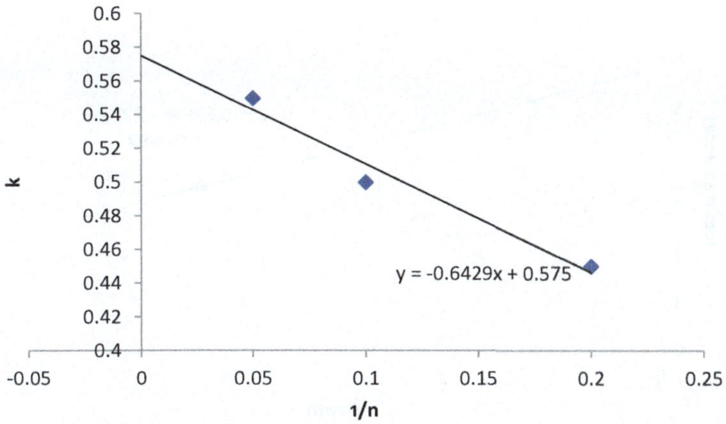

Fig. 4.3 Convergence of k with increasing n or decreasing $1/n$; the converged value is the y-intercept of the fitted straight line

$$I = I_0 \exp(-kx). \tag{4.7}$$

As can be seen, the irradiance is predicted to reduce with depth exponentially. This is known as the Beer–Lambert Law, with k being called the attenuation coefficient.

To find k, we note that the logarithmic form of Eq. (4.7) is a linear equation of $\log_e I$ versus x, with $-k$ being the slope of the graph and $\log_e I_0$ the y-intercept. Figure 4.4 gives such a plot for the experimental data in Fig. 4.2. As can be seen, the data are well fitted with a straight line, confirming the validity of the Beer–Lambert Law. Thus, fundamentally, light loses a constant portion of its intensity every time it passes through a unit thickness of a material.

The attenuation coefficient given by the analytical solution ($k = 0.59$) is also similar to the value of 0.58 given by the numerical solution above. As shown in Fig. 4.5, the numerical predictions with $k = 0.59$ approach the experimental data as the number of divisions increases.

4.5 Specular Reflection

The above discussion neglected the loss of irradiance through specular reflection. As the curing light hits the free surfaces of a resin composite, during either its entering into or exiting from the material, a significant portion of it gets reflected back.

Let r be the portion of curing light that is reflected off from a free surface, so $1 - r$ is the portion of light transmitted through that free surface. The irradiance of light that first enters the resin composite is thus

$$I_0 = I_\infty(1 - r), \tag{4.8}$$

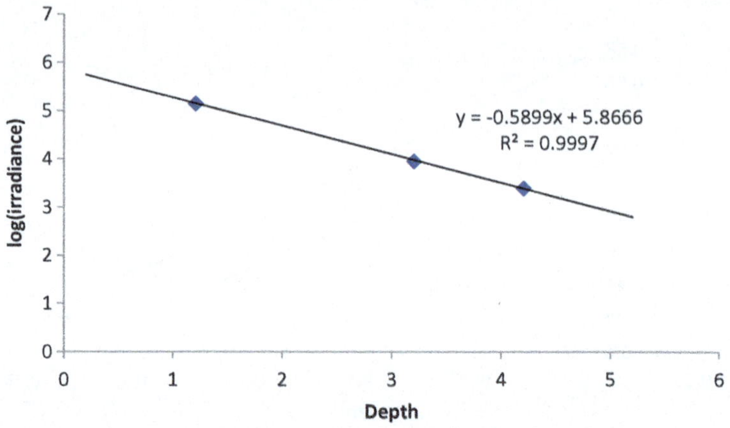

Fig. 4.4 Determination of attenuation coefficient from logarithmic plot of irradiance against depth

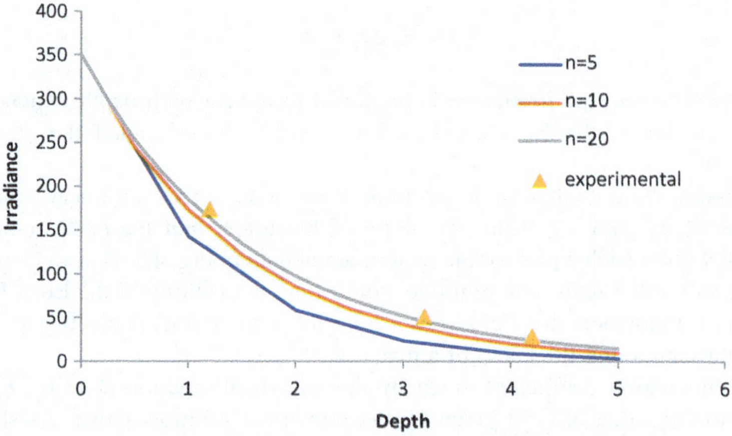

Fig. 4.5 Convergence of numerical results with increasing number of divisions (n); $k = 0.59$

where I_∞ is the full irradiance of light coming from the curing unit. As the light emerges from the other free end of the composite, the same *proportional* loss occurs. Thus, the irradiance we measure coming out of the composite is not the irradiance of light that reaches the end of the material, but rather $(1 - r)$ of it. Introducing these corrections for specular reflection to Eq. (4.7), we get

$$\frac{I}{(1-r)} = I_\infty(1-r)\exp(-kx),$$

$$I = I_\infty(1-r)^2\exp(-kx),$$

$$\log_e I = -kx + \log_e\left[I_\infty(1 - r)^2\right]. \tag{4.9}$$

The fractional loss, r, through specular reflection can, therefore, be obtained from the y-intercept of Fig. 4.4, provided we know I_∞, the full irradiance of the curing light, independently. Note that incorporation of specular reflection into our equation does not change the slope of the $\log_e I$ versus x graph; it only affects the y-intercept. Therefore, the attenuation coefficient can still be determined as above.

4.6 Concluding Remarks

In the above analysis, the attenuation coefficient of the resin composite was assumed to be a constant. In reality, its value changes as the resin composite cures. One of the reasons for the change in light attenuation is the consumption during cure of the initiator molecules that absorb photons to create radicals for propagating polymerization (see Chap. 9). As a result, more light can be transmitted through the material. Another reason is the direct change in optical properties of the resin with polymerization, thus changing the overall attenuation coefficient of the resin composite.

It was mentioned in the introduction that the time required for adequate light curing of resin composite is roughly inversely proportional to the power of the curing light. Thus, it has been suggested that, to save time, more powerful curing lights could be used. However, this so-called reciprocity law does not always hold [7]. Besides, the increased heat from the curing light, coupled with that from the exothermic polymerization reaction, may damage the underlying pulp. In Chap. 6, we shall see an analysis on the heat generated from resin composites and temperature rise as a result during cure.

References

1. Rathbun, M.A., et al.: Cytotoxicity of a BIS-GMA dental composite before and after leaching in organic solvents. J. Biomed. Mater. Res. **25**(4), 443–457 (1991)
2. Sakaguchi, R.L., Douglas, W.H., Peters, M.C.: Curing light performance and polymerization of composite restorative materials. J. Dent. **20**(3), 183–188 (1992)
3. Leprince, J.G., et al.: Progress in dimethacrylate-based dental composite technology and curing efficiency. Dent. Mater. **29**(2), 139–156 (2013)
4. Ilie, N., Stark, K.: Curing behaviour of high-viscosity bulk-fill composites. J. Dent. **42**(8), 977–985 (2014)
5. Jenkins, F.A., White, H.E.: Vibration and waves. In: Fundamentals of Optics. McGraw-Hill, New York (1976)
6. Guo, J., et al.: Determining the temporal development of dentin-composite bond strength during curing. Dent. Mater. **32**(8), 1007–1018 (2016)
7. Wydra, J.W., et al.: The reciprocity law concerning light dose relationships applied to BisGMA/TEGDMA photopolymers: theoretical analysis and experimental characterization. Dent. Mater. **30**(6), 605–612 (2014)

Chapter 5
Release of Ions or Molecules from Dental Restorations

5.1 Introduction

In Chap. 4, we mentioned that unreacted monomers can leach out from under-cured resin composite restorations, which may accelerate their degradation and pose health risks to their host [1]. On the other hand, the release of certain ions or molecules from restorative materials may be beneficial to the surrounding oral tissues. For example, the release of fluoride ions [2] from a restoration may help prevent the formation of dental plaque and recurrent caries. Another example is the release of hydroxyl ions from pulp-capping material to induce the production of calcific barrier at the pulp–dentin interface [3]. The amount and rate of such ion release are important factors in determining the efficacy and longevity of the materials. Materials with slow but sustained release are naturally preferable.

In this chapter, we will look at problems associated with speed rather than slope, i.e., with time rather than position as the independent variable. However, the numerical procedure for calculating the values of the dependent variable remains similar. Put simply, the change in value of a quantity is calculated by multiplying the speed of change with time passed. Also, instead of boundary conditions, we will be considering initial conditions, i.e., values of the dependent variables at the beginning.

Depicted in Fig. 5.1 is a schematic snapshot of the spatial distribution of the concentration (number of particles per unit volume) of a molecule from the surface of a solid material to the surrounding liquid. There is a thin boundary layer right next to the solid surface over which the concentration drops sharply from its peak value before plateauing to the value in the bulk of the liquid. As time passes, the concentration of the molecule in the solid will reduce while that in the liquid will increase.

The phenomenon of leaching is akin to that of an audience leaving a cinema following the end of a movie. In the beginning, when the outside of the cinema is relatively empty, people would be able to exit the cinema rather quickly. There is,

© Springer Nature Switzerland AG 2020
A. Fok, H. P. Chew, *Mathematical Models for Dental Materials Research*,
https://doi.org/10.1007/978-3-030-37849-3_5

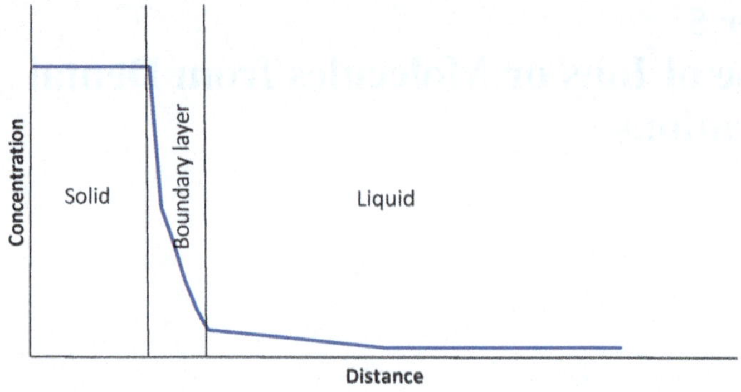

Fig. 5.1 Schematic representation of the spatial distribution of molecular concentration from solid to liquid at a particular time point

thus, an initial spurt of people leaving the cinema. However, some people may hang around in the vicinity either waiting for or chatting to their friends or family members, forming a "boundary layer." As the number of people on the outside of the cinema increases, those (of a reduced number) still inside the cinema will find it harder and thus take longer to exit. At this point, cinema staff may intervene and ask people to move away from the exits quickly, which is similar to stirring of the liquid to reduce the boundary layer.

5.2 Model and Numerical Computation

The thickness of the aforementioned boundary layer depends on the diffusibility of the molecules, which in turn depends on the temperature, motion, and viscosity of the surrounding liquid. Regardless, the boundary layer is usually thin compared to the dimensions of the solid or liquid. As a first approximation, therefore, we can neglect the dimension of this boundary layer and assume that, at a particular moment in time, the rate of leaching is proportional to the difference between the molecular concentration in the solid and that in the bulk of the liquid. Thus, within a short time interval, Δt, we can write the change in molecular concentration in the liquid, C_l, as

$$\Delta C_l = k(C_s - C_l)\Delta t, \tag{5.1}$$

where Δ signifies changes, C_s is molecular concentration in the solid and k is the proportionality or release rate constant. Note that C_s is always greater than C_l, so k is positive as we expect the molecular concentration in the liquid to increase with time. k depends on the surface area of the solid and the mobility of the molecules within both the solid and the liquid.

Starting from the initial values, C_{s0} and C_{l0}, we can calculate the molecular concentration in the liquid at later times as

$$\begin{aligned}
C_{l1} &= C_{l0} + k(C_{s0} - C_{l0})(t_1 - t_0),\\
C_{l2} &= C_{l1} + k(C_{s1} - C_{l1})(t_2 - t_1),\\
C_{l3} &= C_{l2} + k(C_{s2} - C_{l2})(t_3 - t_2),\\
&\vdots\\
C_{ln} &= C_{l(n-1)} + k\big(C_{s(n-1)} - C_{l(n-1)}\big)(t_n - t_{n-1}),
\end{aligned} \tag{5.2}$$

where C_{li} is the molecular concentration in the liquid at time t_i, etc. However, the molecular concentration in the solid, C_{si}, at later times is still unknown. To proceed, we make use of the law of conservation of mass, i.e., what leaves the solid will end up in the liquid without overall loss. Mathematically, this can be expressed as

$$V_s C_s + V_l C_l = \text{constant} = V_s C_{s0} \tag{5.3}$$

for all t, where V_s and V_l are the volumes of the solid and liquid, respectively. The initial molecular concentration in the liquid is assumed to be zero, i.e., $C_{l0} = 0$.

Rearranging Eq. (5.3), we have

$$C_s = \frac{V_s C_{s0} - V_l C_l}{V_s} = C_{s0} - v C_l, \tag{5.4}$$

where $v = \frac{V_l}{V_s}$, i.e., the volume ratio between liquid and solid. Substituting Eq. (5.4) into (5.2) gives:

$$\begin{aligned}
C_{l1} &= C_{l0} + k[C_{s0} - (1+v)C_{l0}](t_1 - t_0),\\
C_{l2} &= C_{l1} + k[C_{s0} - (1+v)C_{l1}](t_2 - t_1),\\
C_{l3} &= C_{l2} + k[C_{s0} - (1+v)C_{l2}](t_3 - t_2),\\
&\vdots\\
C_{ln} &= C_{l(n-1)} + k\big[C_{s0} - (1+v)C_{l(n-1)}\big](t_n - t_{n-1}).
\end{aligned} \tag{5.5}$$

Figure 5.2 shows schematically the molecular concentrations in both the solid and the liquid as time increases. Note the different scales used for the concentration in the solid and that in the liquid—the molecular concentration in the solid is usually much higher than that in the liquid, at least to begin with. Again, the rate constant k can be obtained by trial and error, as described in Chap. 4 for the determination of the attenuation coefficient, using an increasingly small time step to ensure convergence.

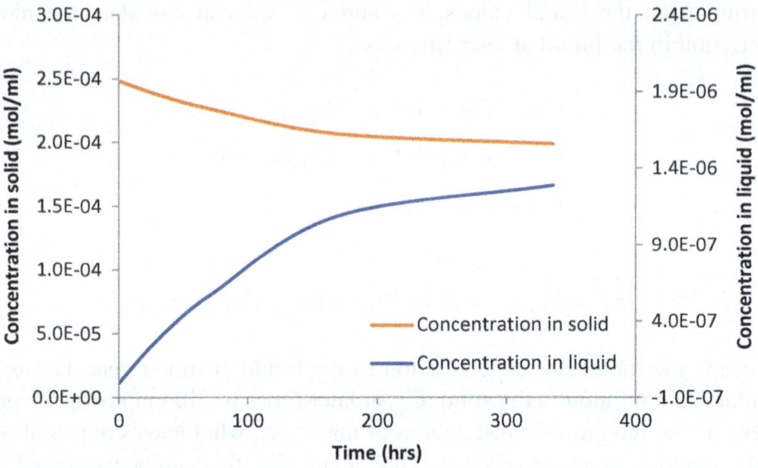

Fig. 5.2 Schematic representations of ion concentration in solid and liquid as time increases

5.3 Analytical Solution

The differential equation derived from Eq. (5.1), after substituting for C_s using Eq. (5.4), can be solved analytically to give:

$$-\left(\frac{1}{1+v}\right)\ln\left[C_{s0}-(1+v)C_l\right]=kt+A, \tag{5.6}$$

where the constant A is found from the initial condition of $C_l = 0$ at $t = 0$. Thus, $A = -\left(\frac{1}{1+v}\right)\ln C_{s0}$ and the solution specific to our problem is

$$C_l=\frac{1}{1+v}C_{s0}\{1-\exp\left[-(1+v)kt\right]\}. \tag{5.7}$$

Using the law of conservation of mass, Eq. (5.4), the molecular concentration in the solid can be shown to be

$$C_s=\frac{1}{1+v}C_{s0}\{1+v\exp\left[-(1+v)kt\right]\}. \tag{5.8}$$

With reference to Eq. (5.6), the release rate constant k can be obtained from the slope of the graph of $\left(\frac{1}{1+v}\right)\ln\left[C_{s0}-(1+v)C_l\right]$ versus t, while the y-intercept is equal to $-\left(\frac{1}{1+v}\right)\ln C_{s0}$.

Figure 5.3 shows the measured fluoride ion concentration in a liquid (deionized water) with a submerged sample of glass–ionomer fissure sealant. This material is commonly used to prevent the formation of caries within the fissures on a tooth surface

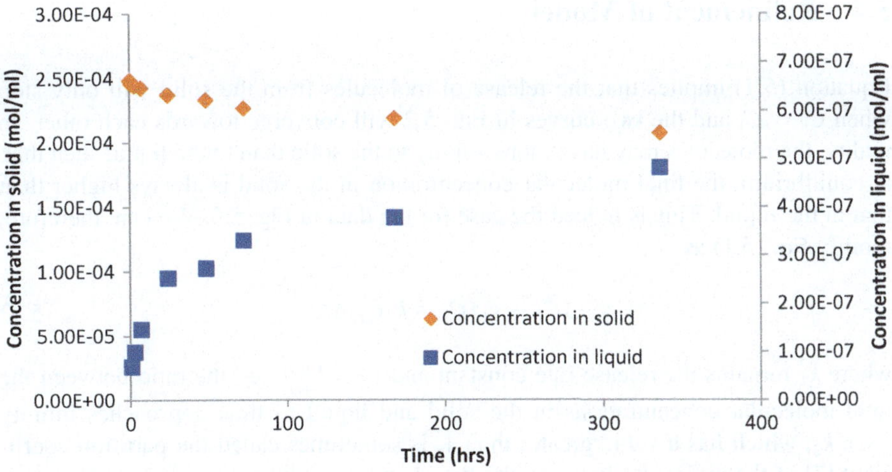

Fig. 5.3 Fluoride ion concentration in a liquid containing a sample of fissure sealant (Unpublished data)

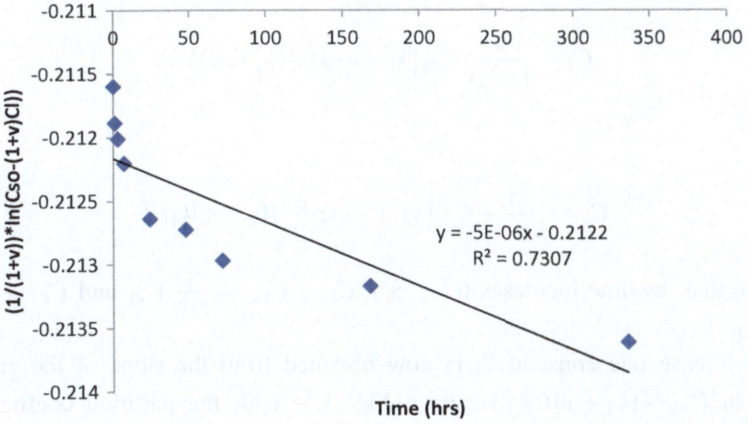

Fig. 5.4 Determination of release rate constant k from fluoride ion concentration data, Fig. 5.3, based on Eq. (5.6)

[4]. The corresponding concentration in the solid sealant is calculated by using the law of conservation of mass, Eq. (5.4). Again, note the different scales used for the concentrations in the solid and the liquid. As can be seen, the trends are very similar to those given by Eqs. (5.7) and (5.8), and shown schematically in Fig. 5.2. Notice the initial burst of ion release. This is typical of many ion-releasing materials [5, 6].

Plotted in Fig. 5.4 is the graph of $\left(\frac{1}{1+v}\right)\ln\left[C_{s0} - (1+v)C_l\right]$ versus t using the data in Fig. 5.3 and the linear regression line fitted to determine the release rate constant k. As can be seen, the model does not fit the data particularly well. It therefore requires some modification.

5.4 Refinement of Model

Equation (5.1) implies that the release of molecules from the solid will only stop when $C_s = C_l$ and the two curves in Fig. 5.2 will converge towards each other. In reality, the molecules may have more affinity to the solid than to the liquid such that, at equilibrium, the final molecular concentration in the solid is always higher than that in the liquid. This is indeed the case for the data in Fig. 5.3. We can, therefore, modify Eq. (5.1) as

$$\Delta C_l = k_1(C_s - k_2 C_l)\Delta t, \tag{5.9}$$

where k_1 remains the release rate constant and $k_2 = \frac{C_{s\infty}}{C_{l\infty}}$, i.e., the ratio between the final molecular concentrations in the solid and liquid as time approaches infinity (∞). k_2, which has a value greater than 1, is sometimes called the partition coefficient [7] of the molecules between the liquid and the solid at equilibrium. Note that when C_s approaches $k_2 C_l$, there will be no further leaching.

The law of conservation of mass still applies, and the solutions for C_l and C_s can be shown to become, respectively,

$$C_l = \frac{1}{k_2 + v} C_{s0}\{1 - \exp[-(k_2 + v)k_1 t]\} \tag{5.10}$$

and

$$C_s = \frac{1}{k_2 + v} C_{s0}\{k_2 + v \exp[-(k_2 + v)k_1 t]\}. \tag{5.11}$$

Note that, as time increases ($t \to \infty$), $C_l \to C_{l\infty} = \frac{1}{k_2+v} C_{s0}$ and $C_s \to C_{s\infty} = \frac{k_2}{k_2+v} C_{s0}$.

The release rate constant k_1 is now obtained from the slope of the graph of $\left(\frac{1}{k_2+v}\right) \ln[C_{s0} - (k_2 + v)C_l]$ versus t, Fig. 5.5, with the partition coefficient k_2 being determined from the ratio of the final molecular concentrations in the solid and liquid in Fig. 5.3. As can be seen by comparing with Fig. 5.4, the data in Fig. 5.5 can now be better fitted with a straight line, justifying the modification of the model.

5.5 Concluding Remarks

We have seen in this chapter how the difference in concentration of an ion or molecule between a restorative material and the surrounding liquid provides the driving force for transfer of material across their interface; see Eqs. (5.1) and (5.9). The rate of transfer reduces with time as the molecular concentrations in the solid

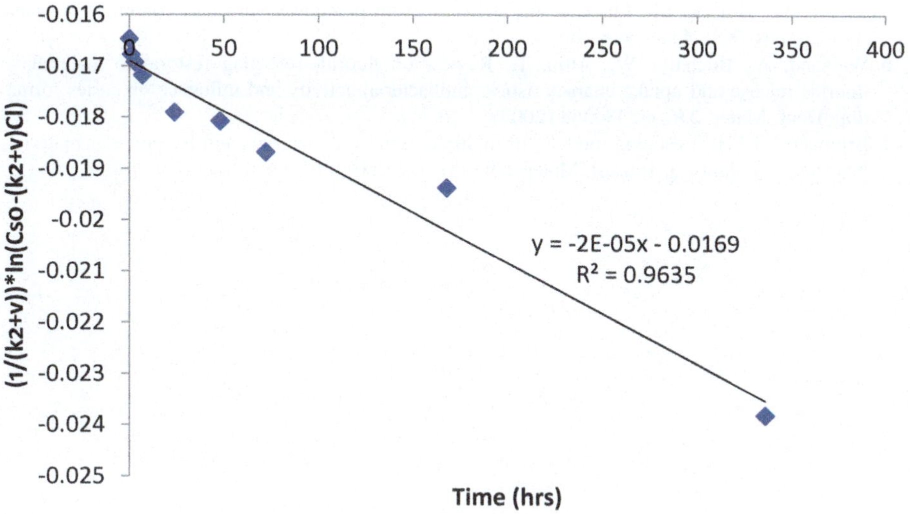

Fig. 5.5 Determination of release rate constant k_1 from fluoride ion concentration data, Fig. 5.3, based on Eq. (5.10)

and liquid approach their equilibrium values. Both the release rate constant and the partition coefficient are material dependent, and they combine to determine the overall rate of leaching.

The release rate constant was assumed to be a constant. In reality, it reduces with time because the driving force (concentration gradient) within the solid itself reduces with time [2]. Over time, it would take longer for the ions or molecules inside the solid to move to the surfaces in contact with the surrounding liquid, thus lowering the rate of material transfer across the interface.

In the next chapter, we will use a similar form for the driving force in terms of temperature to model the heat transfer to the surrounding air from a resin composite under cure. The equations developed will help us determine the amount and rate of heat production by the exothermic polymerization reaction of resin composites.

References

1. Rathbun, M.A., et al.: Cytotoxicity of a BIS-GMA dental composite before and after leaching in organic solvents. J. Biomed. Mater. Res. **25**(4), 443–457 (1991)
2. Hatibovic-Kofman, S., Koch, G.: Fluoride release from glass ionomer cement in vivo and in vitro. Swed. Dent. J. **15**(6), 253–258 (1991)
3. Gandolfi, M.G., et al.: Calcium silicate and calcium hydroxide materials for pulp capping: biointeractivity, porosity, solubility and bioactivity of current formulations. J Appl Biomater Funct Mater. **13**(1), 43–60 (2015)
4. Markovic, D., Peric, T., Petrovic, B.: Glass-ionomer fissure sealants: clinical observations up to 13 years. J. Dent. **79**, 85–89 (2018)

5. Tay, W.M., Braden, M.: Fluoride ion diffusion from polyalkenoate (glass-ionomer) cements. Biomaterials. **9**(5), 454–456 (1988)
6. Wiegand, A., Buchalla, W., Attin, T.: Review on fluoride-releasing restorative materials—fluoride release and uptake characteristics, antibacterial activity and influence on caries formation. Dent. Mater. **23**(3), 343–362 (2007)
7. Brouwers, H.J.H.: Leaching models for multiple immersed materials and for granular materials flushed in a column. J. Hazard. Mater. **53**(1–3), 1–17 (1997)

Chapter 6
Heat Generated from Dental Resin Composites During Curing

6.1 Introduction

In this chapter, we will continue to look at problems associated with speed, i.e. with time as the independent variable. Specifically, we will consider the temperature changes of a light-cured resin composite as a function of time during cure.

A considerable amount of heat is generated by the exothermic polymerization reaction of a dental resin composite, with a reported rise in intrapulpal temperature of 2.9–7.8 °C [1]. A rise in intrapulpal temperature has shown to cause pain and, more importantly, there is concern that it may cause damage to the nerves and soft tissues within the pulp [2, 3] of the restored tooth. If pulp necrosis occurs, the more invasive endodontic procedures would be required. It is therefore essential to measure the exothermic properties of dental resin composites so that the likely temperature rise in the tooth induced by their application can be assessed.

A simple experiment that can be performed to determine the exothermic properties is to measure the temperature rise of a small volume of resin composite during its cure. A thermal couple or an infrared camera can serve the purpose. Figure 6.1 shows the results of such an experiment involving a light-cured resin composite.

Note that there are two stages in the temperature profile shown in Fig. 6.1a of a resin composite under cure. The first stage involves heating by the exothermic reaction and the curing light—the blue curing light is being absorbed by the composite and turned into heat. At the same time, heat is being dissipated by conduction through the surrounding sample holder, convection via the surrounding air and irradiation directly from the free surface of the sample. The second stage involves cooling of the resin composite following the switching off of the curing light, at about 50 s, and termination of the polymerization reaction. Heat loss through the aforementioned mechanisms continues to operate.

The problem of determining the exothermic properties of the resin composite can be much simplified by subtracting the contribution from the curing light, which can be obtained by heating up the cured composite sample for the same duration [4]; see

© Springer Nature Switzerland AG 2020

A. Fok, H. P. Chew, *Mathematical Models for Dental Materials Research*,
https://doi.org/10.1007/978-3-030-37849-3_6

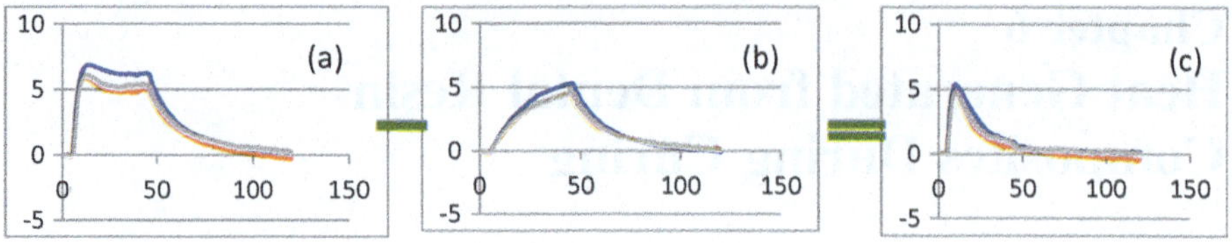

Fig. 6.1 (**a**) Temperature changes (vertical axis, °C) versus time (horizontal axis, *s*) in a light-cured resin composite during and after cure, (**b**) temperature changes in the cured composite heated by the curing light for the same duration, and (**c**) temperature changes in composite due to the exothermic polymerization reaction only

Fig. 6.1b, c. In doing so, we are assuming that the thermal properties of the resin composites are the same before and after cure. It is also assumed that the contributions to the temperature change from the different sources can be superimposed. Mathematically, the latter requires the governing differential equation to be linear [5].

6.2 Model and Numerical Computation

Ignoring contribution from the exothermic polymerization reaction for the time being (Fig. 6.1b), the two main sources for the temperature change in the resin composite are: heat gain from the curing light and heat loss to the surroundings. We will assume that the heat-gain term from the curing light is a constant, which is a function of the curing light's power and the optical and thermal properties of the resin composite. On the other hand, with adequate insulation, heat loss via conduction through the sample-holding mould and irradiation from the free surface are small compared to the heat loss via convection through the surrounding air. The latter is assumed to be proportional to the temperature difference between the sample and the surrounding. This is known as Newton's Law of cooling [6].

Thus, we can write the change in temperature, T, of the sample over a short time interval, Δt, as

$$\Delta T = [k_1 - k_2(T - T_0)]\Delta t, \tag{6.1}$$

where k_1 and k_2 are constants and T_0 is the surrounding room temperature. The first term in the brackets on the right produces the heat gain while the second term produces the heat loss. The cooling rate constant, k_2, is a function of the surface-to-volume ratio, density and specific heat capacity of the resin composite sample. The specific heat capacity of a material is the amount of heat that can be stored in a unit volume per degree of temperature rise.

Note the similarity between the heat-loss term and the ion-release term in Eq. (5.1): the difference in temperature and concentration, respectively, provides the driving force for change in each case. However, the volume of air surrounding the resin composite is sufficiently large to keep the room temperature constant; whereas the liquid containing the leaching solid has a finite volume, so the concentration of ions in the liquid increases with time. Also, Eq. (6.1) contains a heat-gain term, k_1, which was not present in Eq. (5.1). However, if we consider only the cooling of the composite after switching off the curing light, i.e. the second portion of the temperature profile in Fig. 6.1b, then Eq. (6.1) will reduce to a form identical to Eq. (5.1).

Based on Eq. (6.1), recursive equations similar to Eq. (5.2) can be constructed to reproduce the temperature profile as that shown in Fig. 6.1b. Again, by trial and error, we can obtain estimates for the constants k_1 and k_2.

Similarly, if we consider only heating of the cured composite by the exothermic polymerization reaction (Fig. 6.1c), we can write

$$\Delta T = \left[k_3 \mathrm{e}^{-k_4 t} - k_2(T - T_0)\right]\Delta t, \tag{6.2}$$

where k_3 and k_4 are constants. k_3 provides the initial rapid heating by the polymerization reaction, which is assumed to diminish exponentially with a decay rate constant k_4 as the polymerization reaction runs its course. The only difference between Eqs. (6.1) and (6.2) is of course the additional exponential decay function attached to the heat-gain term.

There is no reason why other mathematical functions cannot be used to describe the decay of the exothermic polymerization reaction. However, as shown below, the exponential function allows an analytical solution to be readily obtained for the resulting differential equation.

6.3 Analytical Solutions

Consider first the simpler case of heating the already cured composite with the curing light (Fig. 6.1b). The differential equation for temperature change can be written as

$$\frac{\mathrm{d}T}{\mathrm{d}t} = k_1 - k_2(T - T_0). \tag{6.3}$$

Equation (6.3) can be simplified if we consider temperature relative to the room temperature, i.e. $\theta = T - T_0$. Since T_0 is constant, $\Delta\theta = \Delta T$ and Eq. (6.3) can be simplified as

$$\frac{\mathrm{d}\theta}{\mathrm{d}t} = k_1 - k_2\theta. \tag{6.4}$$

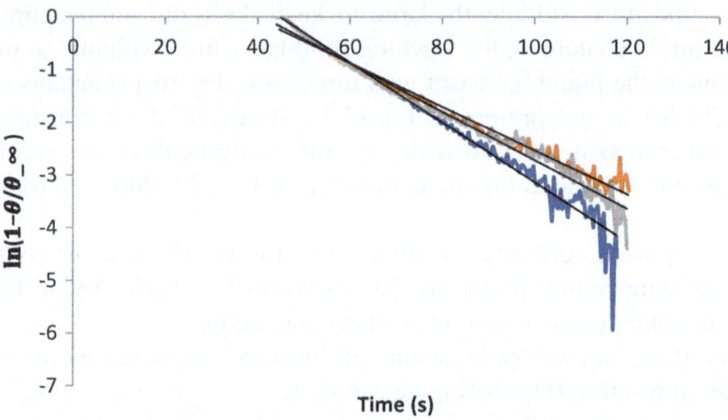

Fig. 6.2 $\ln\left(1-\frac{\theta}{\theta_\infty}\right)$ versus t graph for cured composite heated by curing light

Integrating with respect to t and using the initial condition of $\theta = 0$ (or $T = T_0$) at $t = 0$ give

$$\theta = \frac{k_1}{k_2}\left(1 - \mathrm{e}^{-k_2 t}\right) = \theta_\infty\left(1 - \mathrm{e}^{-k_2 t}\right). \tag{6.5}$$

$\theta_\infty = \frac{k_1}{k_2}$ is the final temperature reached by the already cured composite that is heated by the curing light. It represents the equilibrium state reached when heating by the curing light is balanced by heat loss to the surroundings. The heat-loss or cooling rate constant k_2 can be found from the slope of the graph of $\ln\left(1-\frac{\theta}{\theta_\infty}\right)$ versus t (Fig. 6.2), which will then give the heat-gain term, k_1, from the curing light as $k_1 = k_2\theta_\infty$. The straight lines shown in Fig. 6.2 justifies the use of Newton's law of cooling.

Alternatively, k_2 can be determined from the cooling curve after the curing light is switched off (Fig. 6.1b). For this part of the temperature profile, the solution is again given by solving Eq. (6.4), but without k_1, which will simply be an exponential decay from the maximum temperature reached prior to switching off the curing light.

Next, we consider the more complicated case of the temperature change of the resin composite during curing. After eliminating heating from the curing light, we have from Eq. (6.1)

$$\frac{\mathrm{d}\theta}{\mathrm{d}t} = k_3\mathrm{e}^{-k_4 t} - k_2\theta,$$

$$\frac{\mathrm{d}\theta}{\mathrm{d}t} + k_2\theta = k_3\mathrm{e}^{-k_4 t}. \tag{6.6}$$

For ease of integration, we rewrite the left-hand side of Eq. (6.6) as

$$\frac{1}{e^{k_2 t}}\frac{d}{dt}\left(e^{k_2 t}\cdot\theta\right) = k_3 e^{-k_4 t},$$

$$\frac{d}{dt}\left(e^{k_2 t}\cdot\theta\right) = k_3 e^{(k_2-k_4)t}. \quad (6.7)$$

The equivalence of the two equations can be seen by expanding the left-hand side of the first equation in Eq. (6.7). Integrating with the initial condition of $\theta = 0$ at $t = 0$ gives

$$\theta = \frac{k_3}{k_2 - k_4}\left(e^{-k_4 t} - e^{-k_2 t}\right). \quad (6.8)$$

The temperature profile produced by Eq. (6.8) is shown in Fig. 6.1c. Again, it is the result of a competition between heating from the resin composite (k_3) and cooling via convection (k_2). However, in this case, the heat gain is only transient, the rate of which is controlled by k_4. Note that the temperature change is always positive, irrespective of whether $k_2 > k_4$ or vice versa.

As described above, the heat-loss or cooling rate constant k_2 can be obtained from heating the already cured composite by the curing light only. The maximum rate of heating from the polymerization reaction, k_3, can be obtained from the initial slope of the graph of θ versus t during cure, as $\left.\frac{d\theta}{dt}\right|_{t=0} = k_3$; which is deduced from Eq. (6.6) using the initial condition. To find the remaining unknown, i.e. the decay rate constant, k_4, of the polymerization reaction, we can make use of the maximum temperature θ_c and the time t_c at which it occurs (Fig. 6.1c). At this point, the slope of the curve is zero. Therefore, from Eq. (6.6), $k_3 e^{-k_4 t_c} = k_2\theta_c$. It follows that

$$k_4 = \frac{1}{t_c}\ln\left(\frac{k_3}{k_2\theta_c}\right). \quad (6.9)$$

Finally, by superposition, we can show that the complete solution for the temperature change of the resin composite during cure is

$$\theta = \frac{k_3}{k_2 - k_4}\left(e^{-k_4 t} - e^{-k_2 t}\right) + \frac{k_1}{k_2}\left(1 - e^{-k_2 t}\right). \quad (6.10)$$

It is worth pointing out that the total amount of heat produced by the resin composite is proportional to the sum or integral over time of the pertinent heat-gain term, i.e.

$$\int_0^\infty k_3 e^{-k_4 t}dt = \frac{k_3}{k_4}. \quad (6.11)$$

6.4 Concluding Remarks

Whenever possible, an analytical solution is preferred to a numerical solution, as the former is more general and provides more effective means for determining material parameters. Thus, a more analytical approach is taken in this chapter to illustrate its power.

However, an analytical approach will become unwieldy when the geometry of a problem becomes more complicated. Therefore, to predict the temperature rise in the pulp of a restored tooth, a more detailed computer model such as that based on the finite element method would be required [7].

As expected, the initial rate of heating and the rate of its subsequent decay are closely related to the kinetics of the polymerization reaction of the resin composite. The latter will be explored in detail in Chap. 9.

References

1. Lin, M., et al.: A review of heat transfer in human tooth—experimental characterization and mathematical modeling. Dent. Mater. **26**(6), 501–513 (2010)
2. Zach, L., Cohen, G.: Pulp response to externally applied heat. Oral Surg. Oral Med. Oral Pathol. Oral Radiol. Endod. **19**(4), 515 (1965)
3. Baldissara, P., Catapano, S., Scotti, R.: Clinical and histological evaluation of thermal injury thresholds in human teeth: a preliminary study. J. Oral Rehabil. **24**(11), 791–801 (1997)
4. Kleverlaan, C.J., de Gee, A.J.: Curing efficiency and heat generation of various resin composites cured with high-intensity halogen lights. Eur. J. Oral Sci. **112**(1), 84–88 (2004)
5. Kreyszig, E. (ed.): Advanced Engineering Mathematics, 8th edn. Wiley, Hoboken (1998)
6. O'Sullivan, C.T.: Newton law of cooling – a critical-assessment. Am. J. Phys. **58**(10), 956–960 (1990)
7. Jakubinek, M.B., et al.: Temperature excursions at the pulp-dentin junction during the curing of light-activated dental restorations. Dent. Mater. **24**(11), 1468–1476 (2008)

Chapter 7
Mechanical Failure of Dental Restorations: The Weakest-Link Theory

7.1 Introduction

Tooth tissues and many of the dental restorative materials, e.g. ceramics and resin composites, are brittle in nature [1]. It is typical of these materials to show a large variation in their fracture strength [2]. The latter is attributed to the presence of fracture-causing flaws that have randomly different sizes and orientations [3]. Note that a specimen with the largest flaw is not necessarily the weakest if this flaw is situated in a region or oriented in a direction such that it is not highly stressed. As a result, when testing the load capacity or durability of dental restorations, even when they are anatomically similar, the measured failure loads or lifetimes could exhibit large variations. While the mean strength or life and their standard deviations are still useful comparing different restorative materials, a more mechanistic theory is needed in the design and performance optimization of dental restorations, which can account for their complex shapes and stress distributions. Such analysis is essential before committing to extensive laboratory or even clinical studies, which can be expensive and prone to greater uncertainties and variations associated with, for example, those among a group of patients.

As it is very difficult to eliminate failure completely, a probabilistic approach is needed for analysing dental restorations. Indeed, dentists are highly interested in the rate of failure of a treatment procedure or dental restoration. Thus, introduced in this chapter is the weakest-link theory [3], which has been widely used in conjunction with the Weibull function [4] to predict the failure probability of brittle structures. The use of this theory to analyse and design against failure of dental restorations is presented.

© Springer Nature Switzerland AG 2020

A. Fok, H. P. Chew, *Mathematical Models for Dental Materials Research*,
https://doi.org/10.1007/978-3-030-37849-3_7

7.2 The Weakest-Link Theory

Once a crack forms in a brittle structure under loading, it is very difficult to arrest it. Thus, brittle structures are often likened to a chain of links under tension. When the weakest link of the chain fails, the whole chain fails (Fig. 7.1). Conversely, for the entire chain to survive, all the links must not fail under loading. Likewise, once a crack is initiated in any part of a brittle structure, the latter can be considered to have failed. For the structure to survive, therefore, all parts of it must remain intact during loading.

Mathematically, the weakest-link theory for a chain of n links can be expressed as

$$P_s = p_{s1} \times p_{s2} \times p_{s3} \cdots \times p_{sn}, \tag{7.1}$$

where P_s is the survival probability of the whole chain and p_{si} is the survival probability of the ith link, with $i = 1, 2, 3, \ldots n$.

Assuming that the links are nominally identical in geometry and loading, thus resulting in having the same survival probability, i.e., $p_{s1} = p_{s2} = p_{s3} \cdots = p_{sn} = p_s$, then Eq. (7.1) becomes

$$P_s = p_s^n. \tag{7.2}$$

The relationship between p_s and load σ is usually described by the Weibull function [4]:

$$p_s = \exp\left[-\left(\frac{\sigma}{\sigma_0}\right)^m\right], \tag{7.3}$$

where σ_0 is the characteristic strength of the links and m, the so-called Weibull modulus, is a reciprocal measure of the spread in the strength of the links. The higher m is, the narrower is the spread. For brittle materials, m has a value of around 10. Thus, the survival probability of a brittle material is assumed to decrease exponentially with the applied load.

Fig. 7.1 Failure of a chain is caused by the failure of its weakest link

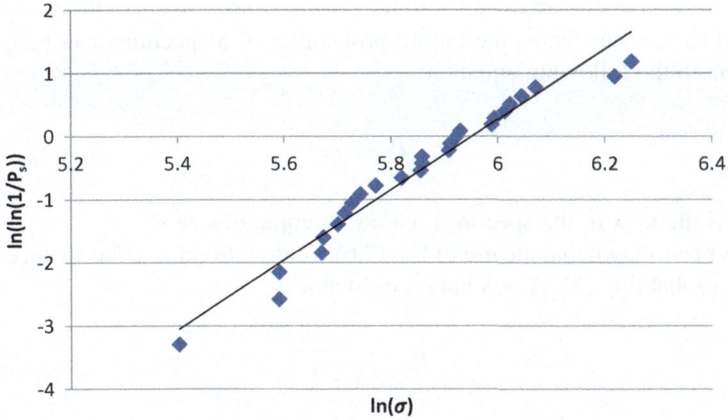

Fig. 7.2 Weibull plot for model composite restorations subjected to fast fracture

Substituting Eq. (7.3) into (7.2) gives

$$p_s = \exp\left[-n\left(\frac{\sigma}{\sigma_0}\right)^m\right]. \tag{7.4}$$

Equation (7.4) can be linearized by taking double logs on both sides to give

$$\ln\ln\left(\frac{1}{P_s}\right) = m\ln(\sigma) + \ln\left(\frac{n}{\sigma_0^m}\right). \tag{7.5}$$

Thus, when $\ln\ln\left(\frac{1}{P_s}\right)$ is plotted against $\ln(\sigma)$, a straight line is expected with m being the slope of the graph and $\ln\left(\frac{n}{\sigma_0^m}\right)$ the y-intercept. This provides a graphical method (see Fig. 7.2) for determining the material properties, m and σ_0, for the brittle restorative material under investigation. In practice, n is replaced by the volume of the specimens and σ_0 can be taken as the mean strength per unit volume of the material.

A corollary of the weakest-link theory is the so-called size effect [3], i.e. the bigger the volume of a specimen, the lower its survival probability under the same load intensity. Intuitively, the bigger is a structure, the more likely it will contain a large failure-causing flaw and, hence, fail at a lower load.

7.3 Estimating Survival Probability

To determine the survival probability of individual specimens tested to failure, the specimens are first ranked in ascending order according to their failure loads. Obviously, if we are to repeat the experiment, when $\sigma = \sigma_n$, the highest failure load, all the specimens are expected to fail; when $\sigma = \sigma_{n-1}$, $\frac{n-1}{n}$ of the specimens are expected to fail, etc. Thus, the failure probability of a specimen can be estimated according to the following equation:

$$P_{\mathrm{f}} = \frac{i}{n}, \tag{7.6}$$

where i is the rank of the specimen within a sample of size n.

It has been shown that the use of Eq. (7.6) would introduce a bias to the estimated survival probability [5]. A less biased estimator is

$$P_{\mathrm{f}} = \frac{i}{n+1}. \tag{7.7}$$

It follows that the survival probability of the specimen is

$$P_{\mathrm{s}} = 1 - P_{\mathrm{f}} = \frac{n+1-i}{n+1}. \tag{7.8}$$

7.4 Application

The failure loads of a batch of root dentin rings "restored" with resin composite and subjected to diametral compression [6] are plotted in Fig. 7.2 according to Eq. (7.5), with the survival probability determined using Eq. (7.8). As can be seen, the data can be well fitted with a straight line, demonstrating the suitability of the weakest-link theory and the Weibull function. As mentioned above, the Weibull modulus can be obtained from the slope of the graph and the characteristic strength from the y-intercept.

Actual dental restorations are complex in shape with nonuniform stress distributions. Again, using the chain as an analogy, imagine each link being subjected to a different load, σ_i. Eq. (7.4) then needs to be modified as

$$\begin{aligned} P_s &= \exp\left[-\left(\frac{\sigma_1}{\sigma_0}\right)^m\right]\exp\left[-\left(\frac{\sigma_2}{\sigma_0}\right)^m\right]\exp\left[-\left(\frac{\sigma_3}{\sigma_0}\right)^m\right]\cdots\exp\left[-\left(\frac{\sigma_n}{\sigma_0}\right)^m\right], \\ &= \exp\left\{-\left[\left(\frac{\sigma_1}{\sigma_0}\right)^m + \left(\frac{\sigma_2}{\sigma_0}\right)^m + \left(\frac{\sigma_3}{\sigma_0}\right)^m \cdots + \left(\frac{\sigma_n}{\sigma_0}\right)^m\right]\right\}, \\ &= \exp\left[-\sum_{i=1}^{n}\left(\frac{\sigma_i}{\sigma_0}\right)^m\right]. \end{aligned} \tag{7.9}$$

In general, a structure under load can be divided into many subunits, small enough that each can be considered to have a different but uniform stress. This is a job well suited for the finite element method which calculates the stresses within a structure by modelling it with elements (springs, essentially) of finite sizes [7]. The elements are connected at nodes where the unknown forces and displacements are used to construct equilibrium equations for the elements, which can be solved using a computer subject to the loading and boundary conditions (physical constraints) assumed. The survival probability of the structure can then be determined by integrating the stress-dependent Weibull function over its entire volume [8]. Thus, provided the material properties, including the Weibull parameters, of the restorative materials are known, the survival probability of a dental restoration of any shape and loading can be predicted using the weakest-link theory, with stresses given by the finite element method. This is a powerful approach that allows new designs of dental restorations to be evaluated effectively prior to actual mechanical testing or clinical trials.

7.5 Fatigue Failure

Clinically, dental restorations are subjected to cyclic loading of a certain amplitude and fail by fatigue after a long period of time or large number of cycles. Extending the weakest-link theory to fatigue failure, Eq. (7.4) can be rewritten as

$$P_s = \exp\left[-\left(\frac{N}{N_0}\right)^{\mu}\right]. \tag{7.10}$$

where N is the number of cycles, N_0 is the characteristic life and μ is the Weibull modulus for fatigue failure. Note that, for simplicity, the volume of the specimens is ignored. Figure 7.3 shows the Weibull plot for dentin specimens failed by cyclic fatigue [4]. The specimens were ranked according to the number of cycles to failure, and their survival probability calculated using Eq. (7.8). As can be seen, the data can again be well fitted with the weakest-link theory in terms of the number of cycles to failure.

The characteristic life N_0 reduces with the load amplitude σ and their relationship is often expressed as a power law [5], i.e.

$$N_0 = \left(\frac{\sigma_0}{\sigma}\right)^{\beta}, \tag{7.11}$$

where σ_0 is the aforementioned characteristic strength for fast fracture and β is the fatigue life reduction factor. Note that when $\sigma = \sigma_0$, the characteristic life is one cycle, as expected.

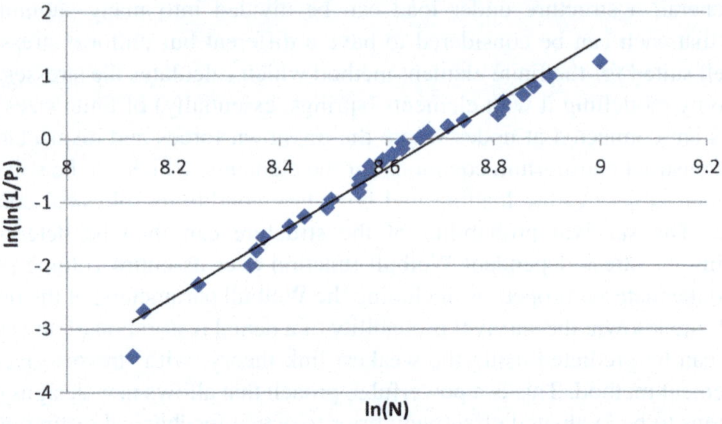

Fig. 7.3 Weibull plot for model composite restorations subjected to cyclic fatigue

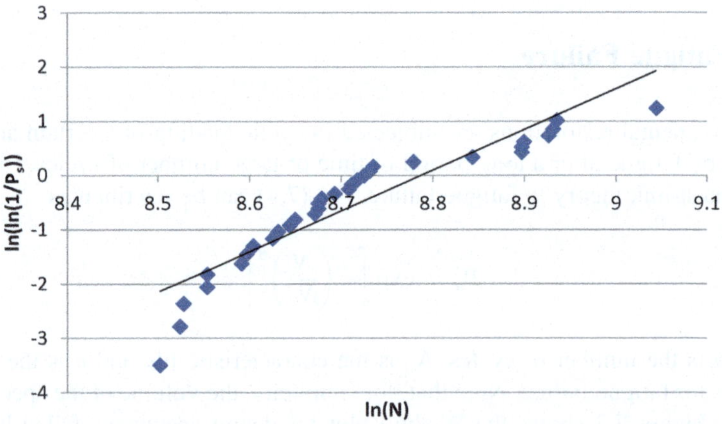

Fig. 7.4 Deviation from an ideal Weibull plot showing possible bimodal behaviour

Substituting Eq. (7.11) into (7.10) gives

$$P_s = \exp\left[-N^{\mu}\left(\frac{\sigma}{\sigma_0}\right)^{\beta\mu}\right]. \tag{7.12}$$

When $N = 1$, i.e. when subjecting the specimens to fast fracture, Eq. (7.12) should reduce to Eq. (7.4). Thus, by comparing the two equations, we can see that

$$m = \beta\mu, \tag{7.13}$$

i.e. the static and fatigue Weibull moduli are linked via a simple linear relationship, with the fatigue life reduction factor β being the constant of proportionality.

In practice, deviations from Eq. (7.5) can often be seen. One of the main reasons is the presence of a threshold stressor number of cycles, below which no failure could occur. In this case, a three-parameter Weibull function that includes the threshold value would be required. Another reason is the emergence of a different failure mode, e.g. interfacial debonding rather than cohesive fracture, which is dominant at one end of the load or fatigue–life range (Fig. 7.4). Post-failure analysis would then be useful in ascertaining the mode of failure for each specimen.

References

1. Park, S., et al.: On the brittleness of enamel and selected dental materials. Dent. Mater. **24**(11), 1477–1485 (2008)
2. Turon-Vinas, M., Anglada, M.: Strength and fracture toughness of zirconia dental ceramics. Dent. Mater. **34**(3), 365–375 (2018)
3. Ashby, M.F., Jones, D.R.H.: Engineering Materials 2, 2nd edn. Butterworth-Heinemann, Oxford (1998)
4. Weibull, W.: A statistical distribution function of wide applicability. J. Appl. Mech. **18**(3), 293–297 (1951)
5. Bergman, B.: How to estimate Weibull parameters. Proc. Br. Ceram. Soc. **39**, 175–185 (1987)
6. Li, K., et al.: Accelerated fatigue testing of dentin-composite bond with continuously increasing load. Dent. Mater. **33**(6), 681–689 (2017)
7. Benham, P.P., Crawford, R.J., Armstrong, C.G.: Mechanics of Engineering Materials, 2nd edn. Longman, Harlow (1996)
8. Smart, J.: The determination of failure probability using Weibull probability-statistics and the finite-element method. Res. Mech. **31**(2), 205–219 (1990)

Chapter 8
Stability of Dental Implants

8.1 Introduction

When dental implants are used to replace missing teeth, it takes time for the surrounding bone to remodel and for the implants to osseointegrate after placement surgery. Only when they are sufficiently integrated with the surrounding bone, will the abutments and prostheses be attached to the implants to restore masticatory functions. Premature loading of implants can lead to their early failure. However, more often than not, there are aesthetics and function necessity to accelerate the placement of the abutments and prosthesis after the implant is being placed. Much has been explored in the quest for the optimum or earliest timing of abutment and prosthesis placement, leading to recommendations such as immediate loading, basal immediate loading, and delayed loading [1, 2].

The density of the thin layer of new bone formed around the surface of an implant is low to begin with, and gradually increases as healing progresses. A method that can be used to monitor the degree of healing is to measure the natural frequency of free vibration of the implant. As the interfacial bone density increases, so will the natural frequency of vibration of the implant. In practice, to overcome damping, which can hinder measurement by quickly reducing the amplitude of free vibration, the implant will be vibrated with a small cyclic force through a transducer attached to its top [3]. The frequency of the applied force will increase gradually with time, and when it reaches the natural frequency of the implant, the amplitude of vibration will reach a maximum. Further increase in the applied force frequency will lead to a reduction in vibration amplitude. This is how the implant's natural frequency of vibration is identified.

In this chapter, we will revisit Newtonian mechanics, first introduced in Chap. 2, and extend its application to dynamic problems. Elementary structural models are used to evaluate the physical and mechanical parameters of materials.

© Springer Nature Switzerland AG 2020

A. Fok, H. P. Chew, *Mathematical Models for Dental Materials Research*,
https://doi.org/10.1007/978-3-030-37849-3_8

8.2 Mass-Spring System

To understand how the interfacial bone density affects the natural frequency of vibration of an implant, let us first consider the free oscillation (damping ignored) of a mass, *m*, connected to a spring of stiffness *k* (force/displacement); see Fig. 8.1. When the mass is pulled or pushed along the spring's axis from its equilibrium (zero-force) position and then released, it will oscillate with a frequency that depends on *m* and *k*. Note that the force from the spring always opposes the movement of the mass and is proportional to its displacement, *x*. Thus, using Newton's Second Law of Motion, which states that the force applied on a body is equal to its mass times acceleration, we can write

$$m\frac{\mathrm{d}^2x}{\mathrm{d}t^2} = -kx,$$

or

$$\frac{\mathrm{d}^2x}{\mathrm{d}t^2} + \frac{k}{m}x = 0. \qquad (8.1)$$

This is a second-order differential equation as it involves the second derivative of the displacement *x*, i.e. acceleration. To solve it numerically, we can rewrite it into two first-order differential equations by introducing the velocity *v*, i.e. the first derivative of *x*, as a second variable. Thus,

$$\frac{\mathrm{d}x}{\mathrm{d}t} = v,$$

and from Eq. (8.1),

$$\frac{\mathrm{d}v}{\mathrm{d}t} + \frac{k}{m}x = 0. \qquad (8.2)$$

The problem can now be solved by using the following expressions for changes in the two variables, *x* and *v*,

$$\Delta x = v\Delta t,$$

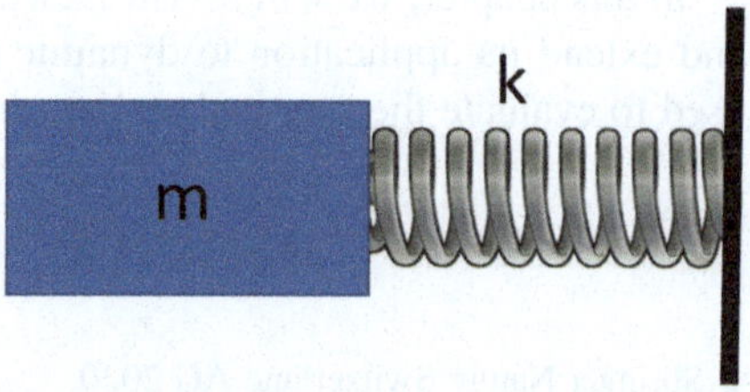

Fig. 8.1 A mass-spring system

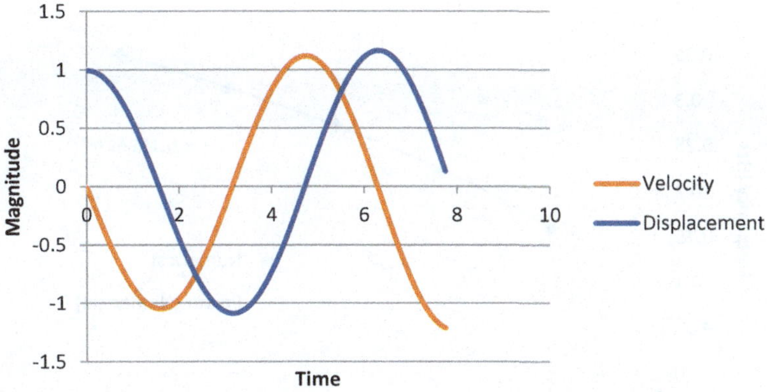

Fig. 8.2 Numerical solutions of displacement and velocity of a mass-spring system undergoing free oscillation

$$\Delta v = -\frac{k}{m} x \Delta t, \tag{8.3}$$

together with the initial conditions $x = X$ and $v = 0$ at $t = 0$, where X is the initial displacement. Note how the two variables depend on each other.

Figure 8.2 shows the computed results for the displacement and velocity of the mass as functions of time, with $X = k/m = 1$. As can be seen, the oscillation appears to be sinusoidal. The increasing amplitudes are due to cumulative errors caused by the finite time steps used in computation, which can be reduced by using smaller time steps. The frequency of oscillation, however, is not very sensitive to the size of the time steps.

By varying the spring stiffness-to-mass ratio, k/m, we can obtain the following results for the frequency of oscillation (Fig. 8.3):

Curve fitting of the numerical data in Fig. 8.3 using a power function indicates that the *natural frequency, f, is proportional to roughly the square root of k/m*. This will be confirmed by considering the analytical solution for Eq. (8.1).

8.3 Analytical Solution

Given that the oscillation of the mass appears to be sinusoidal, it is reasonable to assume that

$$x = X \cos \omega t, \tag{8.4}$$

where the angular frequency ω (rad/s) $=2\pi f$. Note that Eq. (8.4) satisfies the initial condition $x = X$ at $t = 0$.[1]

[1] Mathematically, a solution of a differential equation must satisfy both the equation and the initial or boundary conditions.

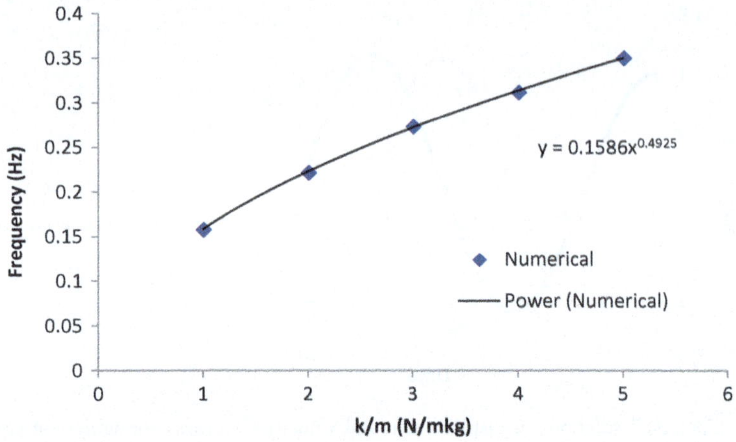

Fig. 8.3 Computed natural frequency of oscillation of the mass-spring system as a function of *k/m*

Differentiating Eq. (8.4) with respect to time gives

$$v = \frac{dx}{dt} = -\omega X \sin \omega t \tag{8.5}$$

which satisfies the other initial condition $v = 0$ at $t = 0$. Differentiating again with respect to time gives:

$$\frac{dv}{dt} = -\omega^2 X \cos \omega t \tag{8.6}$$

Substituting Eqs. (8.4) and (8.6) into (8.1) gives

$$-\omega^2 X \cos \omega t = -\frac{k}{m} X \cos \omega t,$$

which gives

$$\omega = \sqrt{\frac{k}{m}} \text{ or } f = \frac{1}{2\pi}\sqrt{\frac{k}{m}}. \tag{8.7}$$

This confirms the above numerical solution that the *natural frequency of free oscillation increases with the square root of the spring stiffness*. Moreover, it shows that f reduces with the square root of the mass, which agrees with our daily experience that the heavier an object, the slower it oscillates.

8.4 Model for a Vibrating Implant

An implant and its surrounding bone can be simplified as a mass-spring system, with the implant acting as a rigid mass and the bone as a series of elastic springs (Fig. 8.4). The stiffness of the bone depends on its density. For simplicity, we will limit ourselves to the case of in-plane vibration.

Consider first the thin layer of new bone immediately in contact with the implant. Further, consider a small sector of this layer lying at an angle θ from the horizontal axis and subtending a small angle dθ from the centre of the implant (Fig. 8.4a). Given that the axial stiffness of a bar is proportional to its cross-sectional area and Young's modulus (intrinsic stiffness of a material), and inversely proportional to its length [4], the stiffness of the small sector of new bone in the radial direction is

$$\mathrm{d}k_\theta = \frac{r_\mathrm{i}\mathrm{d}\theta.l.E}{t}, \tag{8.8}$$

where r_i is the mean radius, l the out-of-plane depth, E Young's modulus, and t the thickness of this thin layer of new bone.

When the implant moves by a distance X in the horizontal direction, it will produce a radial displacement of $X\cos\theta$ at angle θ, which will induce a radial reaction force from the sector of new bone of

$$\mathrm{d}F_\mathrm{r} = \mathrm{d}k_\theta X\cos\theta = \frac{r_\mathrm{i}\mathrm{d}\theta.l.E}{t}X\cos\theta. \tag{8.9}$$

The horizontal component of this force is

$$\mathrm{d}F_\mathrm{x} = dF_\mathrm{r}\cos\theta = \frac{r_\mathrm{i}\mathrm{d}\theta.l.E}{t}X\cos^2\theta. \tag{8.10}$$

The total horizontal force exerted by the entire layer of new bone is obtained by integrating (8.10) with respect to the angle θ from 0 to 2π, i.e.

$$F_\mathrm{x} = X\frac{r_\mathrm{i}lE}{t}\int_0^{2\pi}\cos^2\theta\mathrm{d}\theta = X\frac{\pi r_\mathrm{i}lE}{t}. \tag{8.11}$$

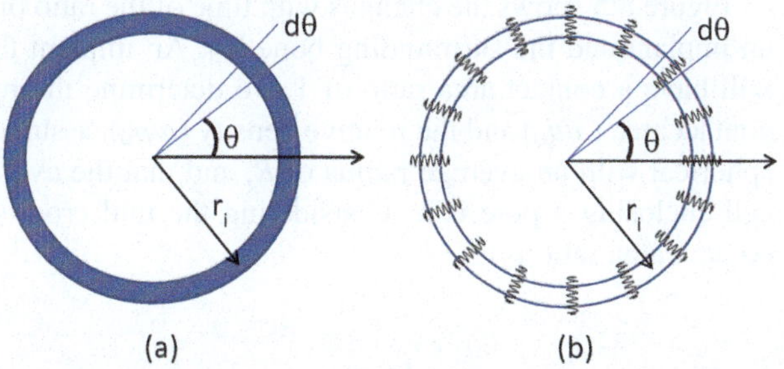

Fig. 8.4 Implant surrounded by (**a**) a thin layer of new bone modelled as (**b**) a series of springs in parallel

Note that, by symmetry, the total vertical force from the top half of the bone as a result of the horizontal displacement X cancels that from the bottom half. This means that, as expected, there will be no vertical motion induced by the initial horizontal displacement.

By definition, the effective stiffness of the thin layer of new bone against horizontal motions is

$$k_{\mathrm{n}} = \frac{F_{\mathrm{x}}}{X} = \frac{\pi r_{\mathrm{i}} l E}{t}. \tag{8.12}$$

The selection of the horizontal axis as the direction of motion is arbitrary. Expression (8.12), therefore, applies to in-plane linear motions along any diameter of the implant. Substituting it and the mass of the implant into Eq. (8.7) will provide an estimate of the system's natural frequency of vibration.

8.5 Effect of Bone Density

The Young's modulus of a material is related to its density (ρ) via [5].

$$E = E_0 \left(\frac{\rho}{\rho_0}\right)^3, \tag{8.13}$$

where E_0 and ρ_0 are the maximum Young's modulus and maximum density of the bone, respectively. Substituting Eqs. (8.13) into (8.12) and then (8.7) gives

$$\frac{\pi r_{\mathrm{i}} l}{t} E_0 \left(\frac{\rho}{\rho_0}\right)^3 = 4 m \pi^2 f^2,$$

which, upon rearrangement, gives

$$f = \left(\frac{r_{\mathrm{i}} l E_0}{4 \pi m t}\right)^{1/2} \left(\frac{\rho}{\rho_0}\right)^{3/2}. \tag{8.14}$$

Figure 8.5 shows the changes with time of the ratio of contact area (a/a_0) between an implant and the surrounding bone [6]. An implant that is fully covered by bone will have a contact area ratio of 1. To determine the relationship between ratio of contact area (a/a_0) and the relative density (ρ/ρ_0), assume that the bone porosities are spherical with an average radius of R, and that the average side length of the cubic cell enclosing a pore is L. Considering the mid cross-section of this unit cell, the contact area ratio is

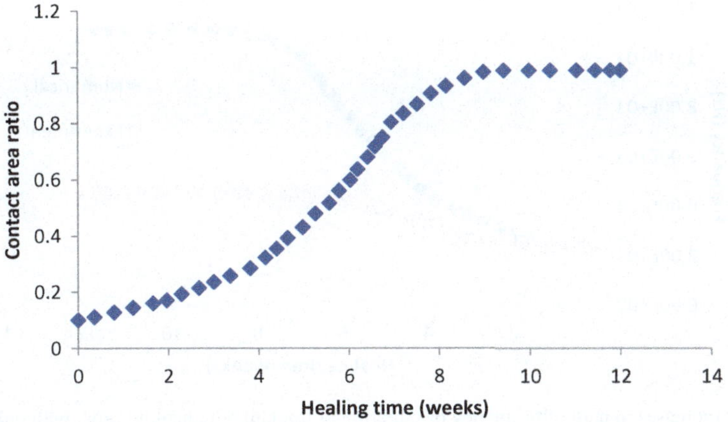

Fig. 8.5 Implant–bone contact area ratio versus healing time [6]

$$\frac{a}{a_0} = \frac{L^2 - \pi R^2}{L^2} = 1 - \pi \left(\frac{R}{L}\right)^2,$$

giving

$$\frac{R}{L} = \left[\frac{1}{\pi}\left(1 - \frac{a}{a_0}\right)\right]^{1/2}. \tag{8.15}$$

The relative density is thus

$$\frac{\rho}{\rho_0} = \frac{L^3 - \frac{4}{3}\pi R^3}{L^3} = 1 - \frac{4}{3}\pi \left(\frac{R}{L}\right)^3 = 1 - \frac{4}{3}\pi \left[\frac{1}{\pi}\left(1 - \frac{a}{a_0}\right)\right]^{3/2} \tag{8.16}$$

Using Eqs. (8.14) and (8.16), the predicted natural frequency of the implant as a function of healing time is plotted in Fig. 8.6. The values assumed for the various parameters are: $E_0 = 0.1$ GPa for trabecular bone, $t = 0.625$ mm, $r_i = 2.5$ mm, $l = 2$ mm, and $m = 0.5$ g. The numerical results are compared with the experimental values given in [6]. As can be seen, although the trends are similar, the agreement between prediction and experiment is poor.

Equation (8.14) is only applicable when the healing bone is soft relative to the adjacent undisturbed bone, i.e. during the early stage of healing. When the stiffness of the new bone approaches that of the adjacent undisturbed bone, we will need to consider the latter's deformation under load as well. Otherwise, as shown in Fig. 8.6, the frequency of vibration of the implant will be overestimated.

To estimate the stiffness of the adjacent undisturbed bone, we simplify it as a thick-walled cylinder (Fig. 8.7). We then divide it into a series of concentric thin annuli with thickness dr.

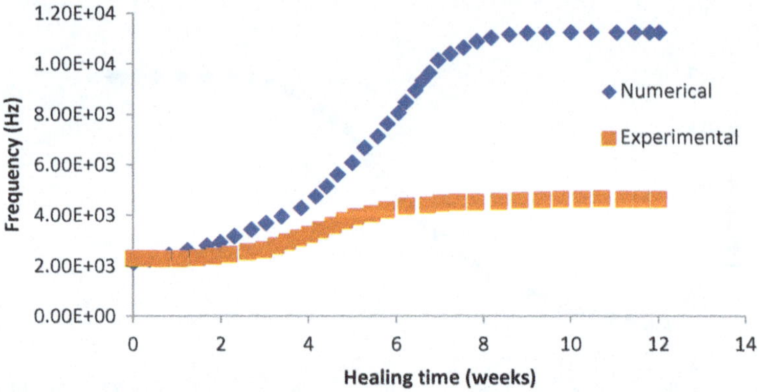

Fig. 8.6 Increases in natural frequency of vibration of implant with healing time, with only healing bone considered

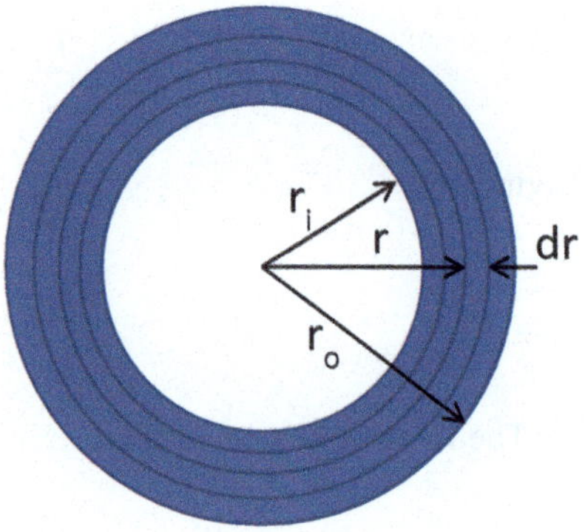

Fig. 8.7 Undisturbed bone around implant modelled as a series of concentric thin annuli

Using the previous result for the healing bone, an annulus of thickness dr at radius r has a stiffness of

$$\mathrm{d}k_{\mathrm{u}} = \frac{\pi r l E_0}{\mathrm{d}r}. \tag{8.17}$$

In contrast to the sectors that are located in parallel within the ring of healing bone (Fig. 8.4), the annuli of the adjacent undisturbed bone are arranged in series. Note that the combined stiffness of springs arranged in parallel is the sum of the stiffnesses of all springs. On the other hand, for springs arranged in series, the same is true for the compliance (displacement/force), which is the inverse of stiffness. Thus, the total stiffness of the adjacent undisturbed bone, k_{u}, is determined by first summing the compliances of all the annuli to determine the total compliance, i.e.

$$\frac{1}{k_{\mathrm{u}}} = \int_{r_{\mathrm{i}}}^{r_o} \frac{1}{\mathrm{d}k_{\mathrm{u}}} = \int_{r_{\mathrm{i}}}^{r_o} \frac{\mathrm{d}r}{\pi r l E_0} = \frac{\ln\left(\frac{r_o}{r_{\mathrm{i}}}\right)}{\pi l E_0}. \tag{8.18}$$

Note that we can use Eq. (8.18) to derive the stiffness of the thin layer of new bone, i.e.

$$k_n = \frac{\pi l E}{\ln\left(\frac{r+t}{r_i}\right)} \approx \frac{\pi r_i l E}{t},$$

when t is small. This is the same as Eq. (8.12). Consideration of special cases using the general solution is useful in ensuring consistency of the solutions.

We can now combine the stiffness of the undisturbed bone with that of the healing bone, which is also in series with the former, to give a total stiffness, k. Thus,

$$\frac{1}{k} = \frac{1}{k_p} + \frac{1}{k_n},$$

$$= \frac{\ln\left(\frac{r_o}{r_i}\right)}{\pi l E_0} + \frac{t}{\pi r_i l E},$$

$$= \frac{r_i E \ln\left(\frac{r_o}{r_i}\right) + t E_0}{\pi r_i l E E_0},$$

or

$$k = \frac{\pi r_i l E E_0}{r_i E \ln\left(\frac{r_o}{r_i}\right) + t E_0}. \tag{8.19}$$

Using Eqs. (8.7), (8.13), and (8.19), the predicted resonance frequency of an implant as a function of healing time is plotted in Fig. 8.8. Again, the Young's modulus of trabecular bone is used and r_o is assumed to be 5.0 mm. As can be seen, the predictions are now much closer to the experimental values. Even better predictions could be obtained if the mass of the bone is also considered.

8.6 Concluding Remarks

In reality, an implant will most likely vibrate with a rocking mode under the excitation of the transducer. For more accurate analysis, the three-dimensional geometry of the implant and surrounding bone must, therefore, be considered. Also, the outer layer of cortical bone with a much higher Young's modulus may have to be taken into account. Despite the simplifications, the mass-spring model

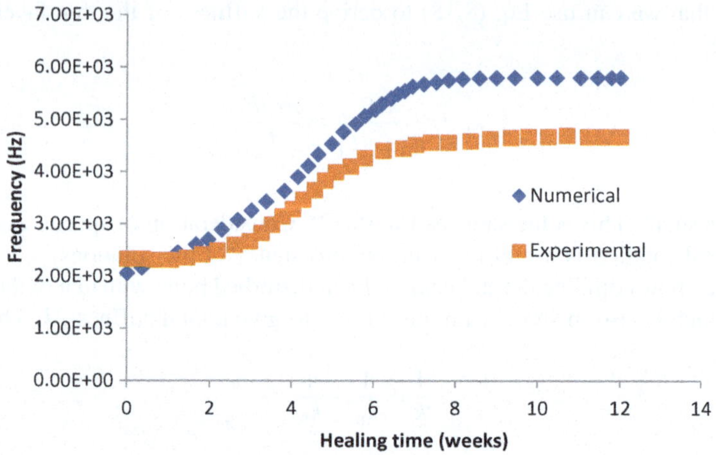

Fig. 8.8 Increases in natural frequency of vibration of implant with healing time, with both healing and undisturbed bone considered

presented in this chapter gives very reasonable predictions for the increase in resonance frequency of an implant during bone healing. In fact, provided a structure can be considered as a rigid body and the stiffness of its surrounding constraints can be reasonably estimated, Eq. (8.7) will provide a good estimate of its natural frequency of vibration.

Also introduced in this chapter is the numerical solution of simultaneous differential equations with more than one variables. Although the computational effort is increased by the increased number of variables, the numerical techniques involved are essentially the same as those for single differential equations. This type of multi-variable problems, which have more complicated but interesting behaviours, will be explored further in the next chapter.

References

1. Garg, R., et al.: Implant survival between endo-osseous dental implants in immediate loading, delayed loading, and basal immediate loading dental implants a 3-year follow-up. Ann Maxillofac Surg. **7**(2), 237–244 (2017)
2. Morton, D., Pollini, A.: Evolution of loading protocols in implant dentistry for partially dentate arches. Periodontol. 2000. **73**(1), 152–177 (2017)
3. Huang, H.M., et al.: Early detection of implant healing process using resonance frequency analysis. Clin. Oral Implants Res. **14**(4), 437–443 (2003)
4. Benham, P.P., Crawford, R.J., Armstrong, C.G.: Mechanics of Engineering Materials, 2nd edn. Longman, Harlow (1996)
5. Huang, H.M., et al.: Resonance frequency assessment of dental implant stability with various bone qualities: a numerical approach. Clin. Oral Implants Res. **13**(1), 65–74 (2002)
6. Perez, M.A., et al.: Computational simulation of dental implant osseointegration through resonance frequency analysis. J. Biomech. **41**(2), 316–325 (2008)

Chapter 9
Kinetics of Photo-Polymerization

9.1 Introduction

In Chap. 6, we looked at the heat produced by the exothermic polymerization reaction of dental resin composites through the temperature rise in a specimen. The concern was that a composite restoration that produced too much heat may damage the underlying pulp. A phenomenological expression (or educated guess) was assumed for the transient heat source term as a function of time (Eq. 6.1). In this chapter, we will look at the mechanism of the polymerization reaction in more detail and derive a mathematical model that considers the various chemical steps involved in converting the monomers into polymers [1]. Such a model would be useful for determining the appropriate compositions of the resin and the intensity of light required to optimize the composite cure. As mentioned before, an inadequately cured composite restoration will have suboptimal clinical performance due to compromised physical and mechanical properties, while the leaching out of uncured monomers may also pose a risk to our health.

We will focus on reactions propagated by radicals, i.e. atoms, molecules, or ions with an unpaired valence electron, created by the curing light. And we will make use of the law of mass action [2] to construct our rate equations. Again, we will be considering simultaneous differential equations of more than one variables, which are usually interdependent. But, as demonstrated in Chap. 8, the numerical techniques used to solve these equations are essentially the same as those used to solve single differential equations discussed in previous chapters. The computation may require more effort as more columns are needed in the spreadsheet, but the mathematics is not much more complicated than previous examples.

© Springer Nature Switzerland AG 2020

A. Fok, H. P. Chew, *Mathematical Models for Dental Materials Research*,
https://doi.org/10.1007/978-3-030-37849-3_9

9.2 The Chemical Reaction

The first step in the photo-polymerization reaction is the creation or *initiation* of primary radicals, $R\bullet$, from the photo-initiators, S, by the curing light. After absorbing light, an initiator molecule will reach a higher electronic state. The excited molecule may (1) give off radiation and/or heat before returning to its ground state, (2) decompose to produce radicals, or (3) react with another substance to produce more complex radicals. Thus,

$$S \leftrightarrow \nu R\bullet, \tag{9.1}$$

where ν is the average number of primary radicals produced by each photo-initiator molecule. Note that the process can be reversible through self-annihilation or recombination of the radicals (see later).

The primary radicals will then react with the resin monomers to form monomer radicals, which then react with more monomers to form polymer chain radicals with increasing length in a process called *propagation*:

$$\begin{aligned} R\bullet + M &\rightarrow P_1\bullet, \\ P_1\bullet + M &\rightarrow P_2\bullet, \\ P_2\bullet + M &\rightarrow P_3\bullet, \\ &\vdots \end{aligned} \tag{9.2}$$

The subscript i of the polymer chain radicals, P_i, indicates the chain length or the number of monomer units in the polymer. Propagation is the step that creates the majority of heat released in the exothermic polymerization process.

A process that happens in parallel with propagation is the annihilation or *termination* of the radicals. This can involve the primary radicals as well as polymer chain radicals of different chain lengths, i.e.

$$\begin{aligned} R\bullet + R\bullet &\rightarrow S, \\ R\bullet + P_1\bullet &\rightarrow P_2, \\ R\bullet + P_2\bullet &\rightarrow P_3, \\ &\vdots \\ P_n\bullet + P_m\bullet &\rightarrow P_{n+m} \quad \text{or} \quad P_n + P_m, \end{aligned} \tag{9.3}$$

where $n, m = 1, 2, 3, \ldots$. If only primary radicals are involved, the process is the reverse of initiation (Eq. 9.1). On the other hand, when two polymeric radicals react, they can form either a longer polymer chain (combination) or two chains of the original lengths (disproportionation). Once terminated, the resulting chains will stop growing. Obviously, if the rate of propagation is higher than that of termination, the reaction will proceed till almost all monomers are used up. Conversely, if the rate of termination is much higher than that of propagation, the resin will end up with short polymers and a significant amount of unreacted monomers.

9.3 The Rate Equations

Next, we will write down rate equations considering the production and removal for each of the above molecules. To do that, we need to invoke the law of mass action [2], which states that the rate of reaction is proportional to the products of the concentrations of all the reactants. Intuitively, the higher the concentration of the reactants, the more frequently they will collide into each other and, hence, the higher the rate of reaction.

9.4 Photo-Initiators and Free Radicals

Since S only reacts with the curing light, the intensity of which is assumed constant, its rate of reduction is simply proportional to its concentration, $[S]$. (Square brackets are used conventionally to indicate concentrations.) At the same time, S can be produced by the self-annihilation or recombination of R. Thus, for the net rate of change of S, we can write

$$\frac{\mathrm{d}[S]}{\mathrm{d}t} = -k_{\mathrm{i}}[S] + \frac{k_{\mathrm{t}}}{\nu}[R\bullet][R\bullet], \tag{9.4}$$

where k_{i} is the initiation rate constant, which increases with the curing light intensity, and k_{t} is the termination rate constant. Note that each dissociation reaction of S produces ν number of $R\bullet$ (Eq. 9.1). Therefore, each reverse termination reaction of $R\bullet$ would produce $1/\nu$ number of S. However, $[R\bullet]$ is usually much smaller than $[S]$ due to the former's high rate of consumption by the ubiquitous monomers and other radicals (see below). Consequently, the second, production term for $[S]$ from the recombination of $R\bullet$ may be ignored without a significant effect on the numerical results, and Eq. (9.4) can be decoupled from the other rate equations below and solved separately. The result would simply be an exponential decay for $[S]$ from its initial value; cf. Eq. (4.2).

After being produced by the initiators, the primary radicals will be consumed by the first step of propagation (Eq. 9.2) and by termination (Eq. 9.3). Thus,

$$\begin{aligned}\frac{\mathrm{d}[R\bullet]}{\mathrm{d}t} &= \nu k_{\mathrm{i}}[S] - k_{\mathrm{p}}[R\bullet][M] - k_{\mathrm{t}}[R\bullet]([R\bullet] + [P_1\bullet] + [P_2\bullet] + [P_3\bullet] + \cdots),\\ &= \nu k_{\mathrm{i}}[S] - k_{\mathrm{p}}[R\bullet][M] - k_{\mathrm{t}}[R\bullet]([R\bullet] + [P\bullet]).\end{aligned} \tag{9.5}$$

where k_p is the propagation rate constant and $[P\bullet]$ the total concentration of all-polymer chain radicals. For simplicity, the rate constant for termination is assumed to be the same for all radicals, including the primary radicals, irrespective of their chain length. In reality, the longer the chain the less mobile a radical becomes and, hence, the lower the rate constant.

9.5 Monomers

For the monomers M, we can write

$$\begin{aligned}\frac{\mathrm{d}[M]}{\mathrm{d}t} &= -k_p[M]([R\bullet] + [P_1\bullet] + [P_2\bullet] + [P_3\bullet] + \cdots),\\ &= -k_p[M]([R\bullet] + [P\bullet]).\end{aligned} \tag{9.6}$$

Note that no monomers are produced during polymerization; they only get consumed by propagation.

9.6 Polymer Chain Radicals

For the polymer chain radicals, we will consider the production and removal of molecules of each increasing chain length, i.e.

$$\begin{aligned}\frac{\mathrm{d}[P_1\bullet]}{\mathrm{d}t} &= k_p[R\bullet][M] - k_p[P_1\bullet][M] - k_t[P_1\bullet]([R\bullet] + [P_1\bullet] + [P_2\bullet] + \cdots),\\ &= k_p[R\bullet][M] - k_p[P_1\bullet][M] - k_t[P_1\bullet]([R\bullet] + [P\bullet]),\\ \frac{\mathrm{d}[P_2\bullet]}{\mathrm{d}t} &= k_p[P_1\bullet][M] - k_p[P_2\bullet][M] - k_t[P_2\bullet]([R\bullet] + [P\bullet]),\\ &\vdots\\ \frac{\mathrm{d}[P_{n-1}\bullet]}{\mathrm{d}t} &= k_p[P_{n-2}\bullet][M] - k_p[P_{n-1}\bullet][M] - k_t[P_{n-1}\bullet]([R\bullet] + [P\bullet]),\\ \frac{\mathrm{d}[P_n\bullet]}{\mathrm{d}t} &= k_p[P_{n-1}\bullet][M] - k_p[P_n\bullet][M] - k_t[P_n\bullet]([R\bullet] + [P\bullet]).\end{aligned} \tag{9.7}$$

where n is the longest chain length. For each chain length, the first term is the production through propagation from chain radicals that are one unit shorter. The second term is the consumption in producing chain radicals that are one unit longer, again through propagation. The third term is the termination through combining with the primary radicals as well as other polymer radicals of different chain lengths.

Adding all the equations in Eq. (9.7) gives

$$\frac{d([P_1\bullet]+[P_2\bullet]+\cdots)}{dt} = k_p[R\bullet][M] - k_p[P_n\bullet][M] - k_t([P_1\bullet]+[P_2\bullet]+\cdots) \times([R\bullet]+[P\bullet]),$$

$$\frac{d[P\bullet]}{dt} = k_p[R\bullet][M] - k_p[P_n\bullet][M] - k_t[P\bullet]([R\bullet]+[P\bullet]). \tag{9.8}$$

Note that in each row the first removal term, through propagation, cancels the production term, also through propagation, in the row below. This leaves us, after adding all the equations in Eq. (9.7), with only the production term from the primary radicals in the first row and the removal term through propagation of $P_n\bullet$ in the final row. Again, for simplicity, it is assumed that the propagation rate constants for all chain lengths are the same. In reality, the longer the chain, the less mobile is the radical and, hence, the lower the rate of reaction and its concentration. Consequently, Eq. (9.8) can be simplified by ignoring the second term on the right-hand side of the equation that involves the longest polymeric radicals, $P_n\bullet$, i.e.

$$\frac{d[P\bullet]}{dt} = k_p[R\bullet][M] - k_t[P\bullet]([R\bullet]+[P\bullet]). \tag{9.9}$$

If we now add Eqs. (9.9) to (9.5), we can see that the propagation terms cancel and we get the following rate equation for the sum of all radicals:

$$\frac{d([R\bullet]+[P\bullet])}{dt} = \nu k_i[S] - k_t([R\bullet]+[P\bullet])([R\bullet]+[P\bullet]),$$

$$= \nu k_i[S] - k_t([R\bullet]+[P\bullet])^2. \tag{9.10}$$

Thus, the rate of change of total radical concentration depends only on their termination and the initiation of primary radicals. It is not sensitive to their propagation.

9.7 Computation

Using the numerical scheme as described in the previous chapters, the concentrations of S, M, R,• and P• can all be calculated at different time points using the initial values of $[M]$ and $[S]$, provided the rate constants are known. However, it is often useful to calculate with normalized concentrations so as to ease comparison between different conditions. For example, by normalizing the concentrations against the initial concentration of the monomer, $[M_0]$, Eq. (9.4) can be rewritten as:

$$\frac{\mathrm{d}\frac{[S]}{[M_0]}}{k_\mathrm{i}\mathrm{d}t} = -\frac{[S]}{[M_0]} + \frac{[M_0]k_\mathrm{t}}{\nu k_\mathrm{i}}\frac{[R\bullet]}{[M_0]}\frac{[R\bullet]}{[M_0]},$$

$$\frac{\mathrm{d}[s]}{\mathrm{d}\tau} = -[s] + k_\mathrm{ti}[r\bullet][r\bullet], \tag{9.11}$$

where $[s] = [S]/[M_0]$, $[r\bullet] = [R\bullet]/[M_0]$, $k_\mathrm{ti} = [M_0]k_\mathrm{t}/\nu k_\mathrm{i}$, and $\tau = k_\mathrm{i}t$. The other rate equations and constants can be modified similarly:

$$\frac{1}{\nu}\frac{\mathrm{d}[r]}{\mathrm{d}\tau} = [s] - k_\mathrm{pi}[r\bullet][m] - k_\mathrm{ti}[r\bullet]([r\bullet] + [p\bullet]), \tag{9.12}$$

$$\frac{1}{\nu}\frac{\mathrm{d}[m]}{\mathrm{d}\tau} = -k_\mathrm{pi}[m]([r\bullet] + [p\bullet]), \tag{9.13}$$

$$\frac{1}{\nu}\frac{\mathrm{d}[p\bullet]}{\mathrm{d}t} = k_\mathrm{pi}[r\bullet][m] - k_\mathrm{ti}[p\bullet]([r\bullet] + [p\bullet]). \tag{9.14}$$

where $[m] = [M]/[M_0]$, $[p\bullet] = [P\bullet]/[M_0]$, and $k_\mathrm{pi} = [M_0]k_\mathrm{p}/\nu k_\mathrm{i}$. Note that all the new parameters are dimensionless.

Figure 9.1 shows plots for the normalized concentrations of S, M, $R\bullet$ and $P\bullet$ using $\nu = k_i = k_p = k_t = 1$. The initial normalized concentration of S is 0.1.

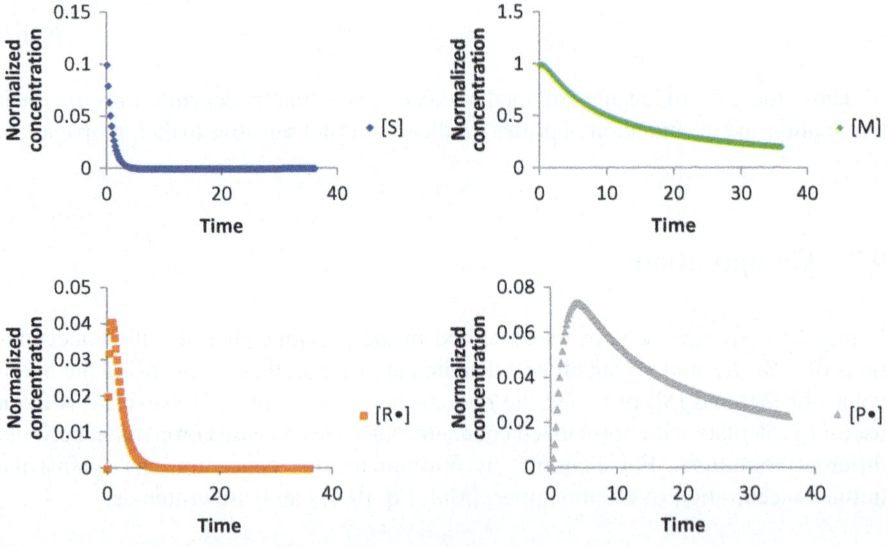

Fig. 9.1 Normalized concentrations of S, M, $R\bullet$ and $P\bullet$ versus time

Note how the initiator and monomer concentrations reduce exponentially with time from their initial values. For the primary and polymeric radicals, there is an initial rapid production followed by an approximate exponential decay.

The degree of conversion of the resin, which indicates the quality of cure, can be calculated from the proportion of monomers consumed, i.e. $(M_0 - M)/M_0$. This is shown in Fig. 9.2. The corresponding rate of conversion is plotted in Fig. 9.3. As can be seen, the rate of conversion has a profile very similar to that of the temperature rise due to the exothermic reaction in Chap. 6.

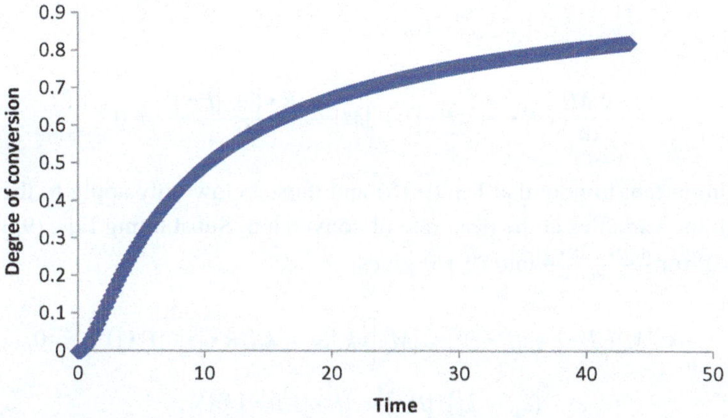

Fig. 9.2 Degree of conversion versus time

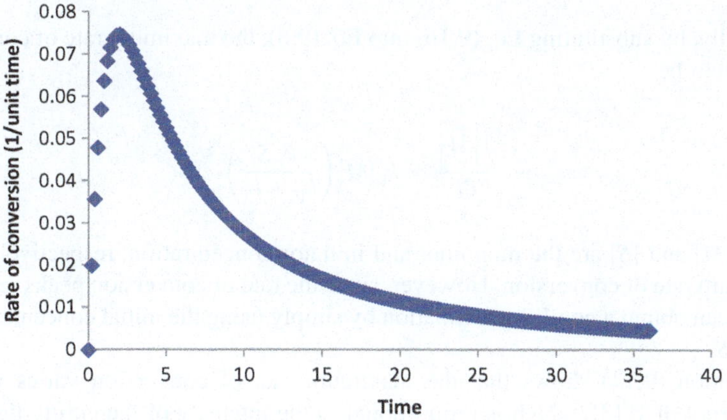

Fig. 9.3 Rate of conversion versus time

9.8 Effect of Light Intensity

Clinically, it would be of benefit to both the dentist and the patient to shorten the time of treatment, and it is suggested that using higher light intensity would reduce the time of cure of the resin composite. We will now examine, analytically, the relationship between the rate of conversion and the rate constant k_i, which is proportional to the intensity of the curing light.

The peak rate of conversion in Fig. 9.3 can be found by making use of the fact that the slope of a curve is zero at a peak. Thus, differentiating Eq. (9.6) with respect to time and equating the result to zero, we get

$$\frac{d^2[M]}{dt^2} = -k_p \frac{d[M]}{dt}([R\bullet] + [P\bullet]) - k_p[M]\frac{d([R\bullet] + [P\bullet])}{dt} = 0,$$

or

$$\frac{d[M]}{dt}([R\bullet] + [P\bullet]) + [M]\frac{d([R\bullet] + [P\bullet])}{dt} = 0. \tag{9.15}$$

It is important to note that Eq. (9.15) and those below only apply to the specific values of the variables at the peak rate of conversion. Substituting Eqs. (9.6) for $\frac{d[M]}{dt}$ and (9.10) for $\frac{d([R\bullet]+[P\bullet])}{dt}$ into (9.15) gives.

$$-k_p[M]([R\bullet] + [P\bullet])^2 + [M]\left[\nu k_i[S] - k_t([R\bullet] + [P\bullet])^2\right] = 0,$$

$$(k_p + k_t)([R\bullet] + [P\bullet])^2 = \nu k_i[S],$$

$$([R\bullet] + [P\bullet]) = \left(\frac{\nu k_i[S]}{k_p + k_t}\right)^{1/2}. \tag{9.16}$$

Finally, by substituting Eq. (9.16) into Eq. (9.6), the maximum rate of conversion is found to be

$$-\frac{d\left[\hat{M}\right]}{dt} = k_p[M]\left(\frac{\nu k_i[S]}{k_p + k_t}\right)^{1/2}, \tag{9.17}$$

where $[M]$ and $[S]$ are the monomer and initiator concentration, respectively, at the maximum rate of conversion. However, since the rate of conversion peaks very early on, we can obtain a good approximation by simply using the initial concentrations of M and S.

Equation (9.14) shows that the maximum rate of conversion varies with the square root of k_i [3], which is proportional to the intensity of the curing light. This is confirmed in Fig. 9.4 that shows a straight-line relationship between the peak rate

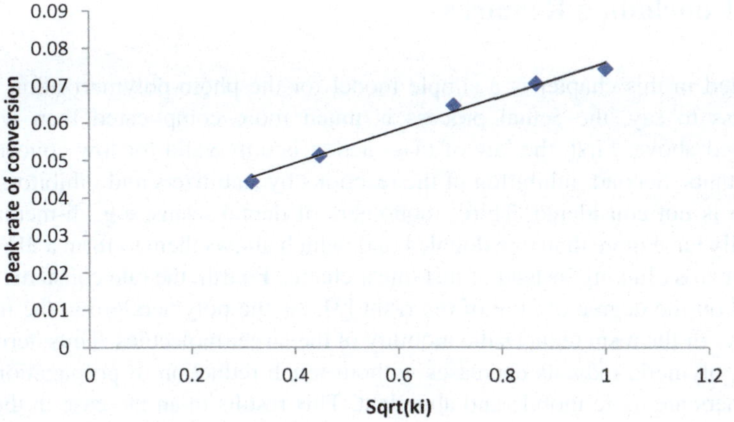

Fig. 9.4 Peak rate of conversion versus $\sqrt{k_i}$

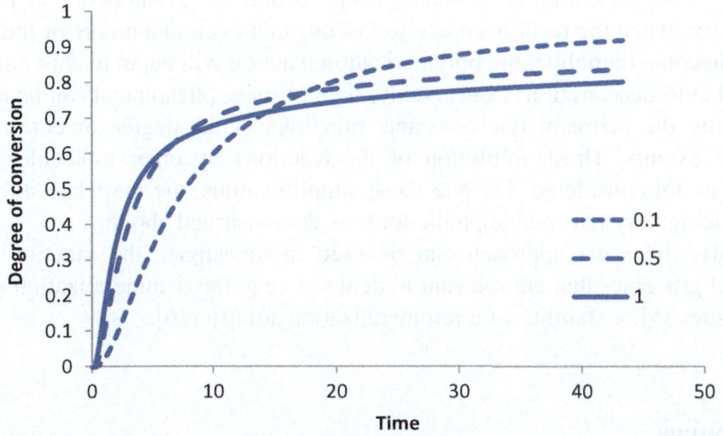

Fig. 9.5 Effect of K_i on the final degree of conversion

of conversion and the square root of k_i. Interestingly, though, increasing the light intensity would reduce the final degree of cure (Fig. 9.5), which has been seen in experiments [4]. This is attributed to the fact that producing primary radicals too quickly, in the beginning, would increase the rate of termination of all radicals (Eq. 9.10). As the amount of initiators is finite, there will be fewer radicals left to sustain the propagation. While using a more powerful curing light may reduce the time of curing, one should be aware of the consequence of a lower final degree of conversion (Fig. 9.5). Increasing the amount of initiator may help alleviate this problem.

9.9 Concluding Remarks

Presented in this chapter is a simple model for the photo-polymerization process. Needless to say, the actual process is much more complicated than what was described above. First, the law of mass action is only valid for low concentrations of reactants. Second, inhibition of the reactions by stabilizers and inhibitors such as oxygen is not considered. Third, monomers of dental resins, e.g. di-methacrylate, generally have more than one double bond, which allows them to form a 3D network through cross-linking instead of just linear chains. Fourth, the rate constants actually depend on the degree of cure of the resin [3]. As the polymers grow, the increased viscosity of the resin reduces the mobility of the larger molecules. Thus, termination of the polymeric radicals decreases without much reduction in propagation, as the monomers are more mobile and abundant. This results in an increase in the rate of conversion or auto-acceleration. In the case of cross-linking polymers, viscosity can increase significantly at very low degrees of conversion, thus limiting the mobility of all molecules early on. However, polymerization is sustained by the diffusion of radicals through the mobile pendant groups within the 3D network of polymers. Eventually, when the resin becomes so viscous that even monomers or the pendant groups become immobile, the polymerization reaction will begin to shut down. This is called auto-deceleration. Conceivably, both of these phenomena can be modelled by making the pertinent rate constants functions of the degree of cure, albeit to different extents. Third, inhibition of the reactions by other molecules such as oxygen is not considered. Despite these simplifications, the model is still capable of producing very reasonable predictions, as demonstrated above.

Finally, the same approach can be used to investigate the kinetics of other chemical processes that are relevant to dentistry, e.g. the demineralization of dental hard tissues [5] or stability of a remineralization solution [6].

References

1. Oster, G., Yang, N.L.: Photopolymerization of vinyl monomers. Chem. Rev. **68**(2), 125 (1968)
2. Fowler, A.C.: Mathematical Models in the Applied Sciences. Cambridge University Press, Cambridge (1997)
3. Goodner, M.D., Bowman, C.N.: Development of a comprehensive free radical photopolymerization model incorporating heat and mass transfer effects in thick films. Chem. Eng. Sci. **57**(5), 887–900 (2002)
4. Wydra, J.W., et al.: The reciprocity law concerning light dose relationships applied to BisGMA/TEGDMA photopolymers: theoretical analysis and experimental characterization. Dent. Mater. **30**(6), 605–612 (2014)
5. Kim, J.L., Cussler, E.L.: Dissolution and reprecipitation in model systems of porous hydroxyapatite. AICHE J. **33**(5), 705–710 (1987)
6. Yun, J.J., et al.: A kinetic model for hydroxyapatite precipitation in mineralizing solutions. Cryst. Growth Des. **18**(5), 2717–2725 (2018)

Index

© Springer Nature Switzerland AG 2020
A. Fok, H. P. Chew, *Mathematical Models for Dental Materials Research*,
https://doi.org/10.1007/978-3-030-37849-3

The manufacturer's authorised representative in the EU is Springer Nature Customer Service Centre GmbH, Europaplatz 3, 69115 Heidelberg, Germany. If you have any concerns regarding our products, please contact ProductSafety@springernature.com

Printed and bound by CPI Group (UK) Ltd, Croydon, CR0 4YY
30/09/2026
02225465-0001